Interactive Wittgenstein

SYNTHESE LIBRARY

STUDIES IN EPISTEMOLOGY, LOGIC, METHODOLOGY, AND PHILOSOPHY OF SCIENCE

VOLUME 349

For further volumes:
http://www.springer.com/series/6607

Interactive Wittgenstein

Essays in Memory of Georg Henrik von Wright

Edited by

Enzo De Pellegrin

Springer

Dr. Enzo De Pellegrin
edp@gmx.at

ISBN 978-1-4020-9908-3 e-ISBN 978-1-4020-9909-0
DOI 10.1007/978-1-4020-9909-0
Springer Dordrecht Heidelberg London New York

Printed on acid-free paper

Springer is part of Springer Science+Business Media (www.springer.com)

Preface

When Georg Henrik von Wright died in his native town of Helsinki on 16 June 2003, aged 87, he left behind a rich and varied legacy of scholarly achievements. Apart from the renown he had attained for his work in philosophical logic and as a quintessential Continental thinker on cultural topics, the polyglot Finn of Scottish descent had earned the esteem of philosophers and scholars around the globe for his wide-ranging efforts to make the philosophical thought of Ludwig Wittgenstein available to an interested public. As a friend of Wittgenstein's, his successor as Professor of Philosophy at Cambridge University and one of the original legatees of his literary estate, von Wright played a crucial part in compiling and organizing the vast body of Wittgenstein's manuscripts and typescripts. This, in itself, was no minor feat, for what is known simply as "the *Nachlass*" among experts turned out to be a geographically scattered set of interrelated texts. In an effort to tame this abundance, von Wright devised a simple classificatory framework and created an annotated catalogue, both of which remain in use by specialists today, serving as indispensable guides to the corpus of Wittgenstein's philosophical writings.

Von Wright was also deeply involved in extracting from the *Nachlass* many of the books through which Wittgenstein's philosophy has reached a broader philosophical audience. Among the volumes he edited and co-edited are *Remarks on the Foundations of Mathematics*, *Notebooks 1914–1916*, *Culture and Value*, *Zettel*, *On Certainty*, *Notes on Logic*, and four volumes of remarks on the philosophy of psychology.

But von Wright not only helped open up a wide new field of inquiry, now known as Wittgenstein Studies, and define its borders through his editorial work. He also helped us place Wittgenstein's work in the context of the philosopher's life by producing an early memoir and by publishing, in collaboration with Brian McGuinness, Wittgenstein's correspondence with, among others, Bertrand Russell, G. E. Moore and Frank Plumpton Ramsey. Moreover, the carefully researched accounts he provided of the tangled textual history of Wittgenstein's early masterpiece, the *Tractatus Logico-Philosophicus*, and the posthumously published *Philosophical Investigations* are considered classics in the field to this day.

Early on, von Wright took strides to make the material basis of his scholarly and editorial work accessible to other specialists. He opened his collection of copies from the *Nachlass* to a countless number of visitors at the philosophy department at the University of Helsinki. Working alongside Norman Malcolm, von Wright supervised the production of a set of microfilms, which, in duplicated form, would remain the primary resource for scholarly research on Wittgenstein's philosophy for several decades until the Wittgenstein Archives at the University of Bergen and Oxford University Press released the results of many years of steadfast efforts to present text and images of the *Nachlass* in machine-readable form on CD-ROM. This venture, too, received crucial support from von Wright.

To honor the memory of Georg Henrik von Wright and his manifold contribution to the study of Wittgenstein's lifework, the present volume assembles essays by experts in the field. Its main title derives from the unique focus many of these essays provide on the philosophical significance of Wittgenstein's real and presumed interactions with—and of his reactions to—other thinkers, including his former self, for the development of his thought.

The first three contributions concern nature and scope of Wittgenstein's philosophical and personal interactions with the German philosopher and logician Gottlob Frege, as related chiefly through letters and cards that Frege had addressed to the young philosopher between the years of 1914 and 1920. It was during this period that Wittgenstein served on the side of the Central Powers during World War I and composed his seminal *Tractatus*. Frege's dispatches to the frontline and his letters to the unpublished author after his return from war and captivity are rendered here in the original German with a modern English translation en face, which was jointly undertaken by the late Burton Dreben and by Juliet Floyd. The one-sided correspondence reflects Frege's admiration for the unflinching soldier's ability to keep up scientific work under conditions of hardship. But it also contains stridently worded criticism of a typescript version of the *Tractatus*, which Frege had received in 1919. In addition, as Floyd, who also provides a prefatory note and an indispensable scholarly apparatus, points out in a companion essay entitled "Interpretive Themes", Frege's letters record his response to criticism Wittgenstein had lodged about Frege's paper "Thoughts" in the half of their correspondence that is now missing. Drawing on her own research on the correspondence between Frege and Wittgenstein, Floyd puts the mutual criticisms into perspectives and speculates on the extent to which these criticisms may have played a part in the abrupt termination of their interaction.

The theme of influence, or lack thereof, is the starting point for Eran Guter's exploration of Wittgenstein's thought on music and musical language. In his essay "A Surrogate for the Soul: Wittgenstein and Schoenberg", Guter challenges a widespread assumption, arguing that Wittgenstein and the Austrian composer Arnold Schoenberg had little in common beyond their shared cultural heritage, overlapping social circles in fin-de-siècle Vienna and the notoriety each of them had attained in his respective domain in the first half of the twentieth century. Guter combines a vivid account of Wittgenstein's aesthetic inclinations and the intellectual influences that may have reinforced them with a survey of various philosophical statements by Wittgenstein on music, musical meaning and language. Guter also attempts to form a Wittgensteinian response to Schoenberg's dodecaphonic language and to answer the question as to why Wittgenstein and Schoenberg arrived at very different ideas about contemporary music and the music of the future.

In his essay "The Crash of the Philosophy of the *Tractatus*: The Testimony of Wittgenstein's Notebooks in October 1929", Jaakko Hintikka describes the central features of what he takes to be Wittgenstein's own reassessment of Tractarian doctrines in the years of 1929 and 1930. Building on conclusions in his earlier interpretive work on Wittgenstein, Hintikka focuses on a small number of entries to Wittgenstein's philosophical diaries and other sources that he takes to reflect a momentous shift in Wittgenstein's conception of language. It is this shift, Hintikka argues, that brings about an abrupt and lasting change in Wittgenstein's thinking and radically transforms his stance on a wide range of philosophical questions. Besides providing a translation of the relevant passages from the *Nachlass*, Hintikka examines some of the methodological issues arising from the shift of 1929.

The shift in thought that David Pears has in mind in his contribution to this volume, which is entitled "Linguistic Regularity" and a different version of which appears in his book "Paradox and Platitude in Wittgenstein's Philosophy" (Oxford University Press, 2006), is of another sort altogether. It pertains to basic differences Pears perceives in the ways Wittgenstein treats philosophical questions about linguistic meaning, and in particular linguistic regularity, in the *Tractatus* and in his later writings, such as the *Philosophical Investigations*. Pears assumes that, in his later work, Wittgenstein comes to reject his Tractarian account of the phenomenon of linguistic regularity. Starting with this assumption, Pears sets out to examine the question as to what replaces Wittgenstein's earlier account of the phenomenon in his later considerations on philosophical uses of the notion of meaning.

Finally, with his essay "On a Remark by Jukundus", Joachim Schulte turns our attention to Wittgenstein's views on religious belief and practice. While doubtful that Wittgenstein produced a self-standing philosophy of religion, Schulte takes the scattered remarks on religion in the *Nachlass*, and in particular those on Christianity, to reflect Wittgenstein's distinctive conception of ethics. The centerpiece in Schulte's survey is his analysis of Wittgenstein's qualified rejection of the Pauline doctrine of election by grace as meaningless. Schulte proposes a new reading of Wittgenstein's claim that this doctrine can be accepted as meaningful on one level of religiosity while being rejected as meaningless on another level. Along the way, Schulte develops an account of Wittgenstein's personal level of religiosity that connects his laconic remarks on this matter with other strands in his thinking.

Georg Henrik von Wright (1916–2003). Reproduced by Courtesy of Anna-Maija Hintikka.

Acknowledgments

For their sage advice and generous assistance, particularly during early stages in the evolution of this volume, I thank Jaakko Hintikka and Robert S. Cohen. Moreover, I owe a debt of gratitude to Hermi Kastner and Reginald E. Harris for the unflagging support and encouragement they both provided throughout the past few years. An anonymous referee whose scrupulous eye detected many an infelicity in an earlier version of the manuscript submitted to the publisher deserves special thanks. I also gratefully acknowledge the unstinting support of Ingrid van Laarhoven of Springer Publishing Company. Sadly, David Pears passed away before this volume reached the printer. His original contribution is included here by kind permission of his widow, Anne Pears.

Contents

Prefatory Note to the Frege-Wittgenstein Correspondence

Juliet Floyd[1]

The historical record concerning Wittgenstein was significantly augmented in June 1988, when around five hundred letters to him, from a variety of correspondents—including Georg Trakl, Bertrand Russell, and many others—were discovered in the store room of a real-estate broker in Vienna. The woman who had been ordered to dispose of the old papers noticed the name "Wittgenstein" and stopped to examine them just before they were shredded.[2] Among this trove, now housed at the Brenner Archives at the University of Innsbruck, were twenty-one letters from Frege to Wittgenstein. The first is dated 11 October 1914, the last 3 April 1920. These form the entire correspondence between them that is presently known still to exist.

None of the letters from Wittgenstein to Frege are thought to have survived the bombing of the Münster library in 1945,[3] where they were deposited

[1] I am pleased to thank members of the Boston University Editorial Studies program seminar, 2005, under the auspices of Archie Burnett and Frances Whistler, as well as Gisela Bengtsson, Enzo De Pellegrin, Mirja Hartimo, Malek Husseini, Allan Janik, Akihiro Kanamori, Wolfgang Kienzler, Andrew Lugg, Brian McGuinness, Jennifer Page, R. D. Schindler, Richard Schmitt, Peter Simons, Christian Thiel, and W.V. Quine for their helpful suggestions and encouragement on the translation. Kienzler provided substantial scholarly help, both with detailed comments on the transcription of the original German text and with his many very helpful suggestions for improving accuracy of the translation. Janik supplied me with continual encouragement, as well as with photocopies of the original letters and preprints of his editorial work now largely published in the electronic edition of Wittgenstein's correspondence. Last but certainly not least, Dr. De Pellegrin has been an unfailingly generous and acute editor of the translation at each stage of its preparation, offering many insightful suggestions about the scholarly presentation of the material.

[2] Reinhard Merkel, "Du wirst am Ende verstanden werden", *Die Zeit*, Dossier, No. 18, 28 April 1989, p. 13.

[3] The evidence for this is not absolutely conclusive, according to Kai F. Wehmeier and Hans-Christoph Schmidt am Busch; see their "The Quest for Frege's Nach-

E. De Pellegrin (ed.), *Interactive Wittgenstein*, Synthese Library 349,
DOI 10.1007/978-1-4020-9909-0_1,

for safekeeping (among Frege's papers) by Heinrich Scholz, a professor who had begun to catalogue and archive Frege's papers in the mid-1930s.[4] A list of the dates and rough contents of the letters from Wittgenstein to Frege did, however, survive in Scholz's papers. These are now housed in the Scholz archive at the University of Münster.[5] Below I have appended a list of the complete chronology of the correspondence between Frege and Wittgenstein as it is now thought to have existed, along with Scholz's annotations and a few supplementary remarks of my own concerning the evidence we have for believing that the letters or cards were written on or around these specific dates. As the reader will see, what remains today forms slightly less than half of the original correspondence.[6]

---

lass", in M. Beaney and E. Reck, eds., *Gottlob Frege: Critical Assessments of Leading Philosophers*, vol. I *Frege's Philosophy in Context* (New York: Routledge, 2005), pp. 54–68 for an argument that it is just possible that further material might turn up.

[4] Scholz (1884–1956) was a logician, philosopher and theologian. Made Professor of Theology in Breslau (1917), and of Philosophy in Kiel (1919) and later Münster (1928), he founded the first Institute of Logic and Mathematics in Münster, where the Frege Archiv is still housed.

[5] Scholz's papers are in the Scholz Archive at the University of Münster (see *www.math.uni-muenster.de/math/inst/logik*); correspondence and notes concerning the Frege-Wittgenstein correspondence are in the Frege-Archive at the University of Münster. A detailed history of Frege's papers is given in Albert Veraart, "Geschichte des wissenschaftlichen Nachlasses Gottlob Freges und seiner Edition. Mit einem Katalog des ursprünglichen Bestands der nachgelassenen Schriften Freges", in Matthias Schirn, ed., *Studies on Frege I: Logic and Philosophy of Mathematics/Studien zu Frege I: Logik und Philosophie der Mathematik* (Stuttgart-Bad Cannstatt: Frommann-Holzboog Verlag, 1976), pp. 49–106 and is also reviewed in the Preface to Frege, *Philosophical and Mathematical Correspondence*, ed. B. McGuinness, trans. Hans Kaal (Chicago: The University of Chicago Press/Oxford: Blackwells, 1980), a partial translation of vol. II of Gottlob Frege, *Nachgelassene Schriften und Wissenschaftlicher Briefwechsel*, eds. G. Gabriel et al. (Hamburg: Felix Meiner Verlag, 1976). As Veraart explains (p. 67), Scholz made three slightly differing annotated lists describing Frege's correspondence, known to scholars as the "Scholz Lists" 1, 2 and 3.

[6] As mentioned in the chronology below, there were letters from Wittgenstein's sister Hermine to Frege; I presume the originals were destroyed in the bombing of the Münster library. Copies survived, published now with the range of correspondence between Hermine and Ludwig on the Intelex CD-rom of Wittgenstein's collected correspondence (see footnote 9 below). See also *Wittgenstein Familienbriefe*, eds. B. McGuinness, M. C. Ascher, O. Pfersmann, *Schriftenreihe der Wittgenstein-Gesellschaft*, vol. 23 (Vienna: Hölder-Pichler-Tempsky Verlag, 1996) and, for a discussion of their bearing on issues surrounding the identification of pre-Tractatus manuscripts and notebooks, Brian McGuinness, "Some Pre-Tractatus Manuscripts", in his *Approaches to Wittgenstein: Collected Papers* (New York: Routledge, 2002), pp. 259–269. The letters from Hermine to Frege concerned Hermine's role as an intermediary between her brother and Frege in

As part of his effort to assemble Frege's scientific correspondence for publication, Scholz wrote to Wittgenstein in Cambridge in April 1936, expressing a hope that Wittgenstein might still possess the letters Frege had written to him. Scholz informed Wittgenstein that his own notes had recorded a number of letters and cards exchanged between Frege and Wittgenstein. He then invited Wittgenstein to donate any letters he might still have to the Frege archive that Scholz was in the process of establishing. Wittgenstein wrote back to Scholz to say that, though he did possess a "few" cards and letters from Frege "[their] contents are ... purely personal and not philosophical" and would have little value for a collection of Frege's writings. In any event, because of their sentimental value he would not like the letters to be placed in Scholz's Frege archive.

Since the final four letters voice Frege's critical reactions to Wittgenstein's *Tractatus* and mention Wittgenstein's criticisms of Frege's essay "Der Gedanke", it is difficult to agree that the letters are wholly without philosophical significance—though gauging that significance is, as I argue in my essay in this volume, hardly an easy thing to do. What is clear is that if Wittgenstein had complied with Scholz's request, it is likely that the letters would have been destroyed in 1945.[7]

What follows is a translation of the twenty-one existing pieces of correspondence from Frege to Wittgenstein, along with the two letters exchanged between Scholz and Wittgenstein about them in 1936.[8] All twenty-three of these pieces of correspondence have appeared in electronic form in the original German, with cross-referencing and editorial commentary, on the CD-ROM of Wittgenstein's correspondence distributed by Intelex in the Past Masters Series.[9] I have followed this edition with regard to the German, though not slavishly. With Enzo De Pellegrin's help, some slight adjustments have been made, such as printing umlauts where Frege actually wrote two vowels together. Whenever a different word has been substituted, I have made a footnote.[10]

---

arranging to mail the manuscript of the *Tractatus* to Frege at the end of the war, as well as her role as an organizer of Wittgenstein's manuscripts. In the chronology of the correspondence below I make note of these letters, but have not translated them in what follows, as they shed no substantial light on the Frege-Wittgenstein correspondence.

[7] This is pointed out by R. Schmitt, in his introduction to his translation of the final four letters from Frege to Wittgenstein; see note 11 below.

[8] The latter two letters are presently housed in the Frege Archive at the University of Münster.

[9] CD-ROM, *Ludwig Wittgenstein: Briefwechsel* (Innsbrucker elektronische Ausgabe 2004), eds. Monika Seekircher, Brian McGuinness, Anton Unterkircher, Allan Janik and Walter Methlagl.

[10] The German letters from Frege to Wittgenstein were first published in an issue of *Grazer Philosophische Studien*, vol. 33/34, *Wittgenstein in Focus—Im Brennpunkt: Wittgenstein*, eds. Brian McGuinness and Rudolf Haller (Amsterdam: Rodopi, 1989), pp. 5–33, but since that time editorial scrutiny of the original

Burton Dreben and I prepared the present translation for publication at the suggestion of G.H. von Wright. During the winter of 1998–1999 we generated a partial draft without footnotes and editorial apparatus of the Frege to Wittgenstein letters, extending and slightly amending earlier translations of mine of the final four letters (those which contain Frege's remarks to Wittgenstein about the *Tractatus*).[11] In July 1999 Burton passed away, leaving me to finish the translation, contextualize it with the 1936 Scholz-Wittgenstein correspondence, and provide an introduction and scholarly notes. I thus bear sole responsibility for the final form of the English, as well as all editorial comments.

Throughout this latter part of the project, I have been guided by the ideals Burton and I shared regarding translation. We strove to convey a sense in English of how the academic German style of Frege sounded, including his manner and tone in addressing Wittgenstein, which is somewhat unusual within the corpus of Frege's writings in virtue of their relationship and the circumstances in which they were corresponding. We therefore stressed concision, readability, and tone in the target language, rather than literalness of meaning from the original German.

---

documents has led scholars to change the date on one of the postcards and to slightly alter some of the German wording, in part thanks to Dr. De Pellegrin's able scrutiny of the original German handwriting. Wittgenstein's reply to Scholz was first published in Veraart (op. cit.), p. 106, and excerpted from in Gabriel, et al., ed., *Wissenschaftlicher Briefwechsel* (op. cit.), p. 265.

[11] See Juliet Floyd, "The Uncaptive Eye: Solipsism in the Tractatus" in L. Rouner, eds., *Loneliness* (Notre Dame: Boston Studies in the Philosophy of Religion, 1998), pp. 79–108. The commentary provided in this essay is purely philosophical, and directed solely at the topic mentioned in the essay's title. Another, independent translation of just these four letters was prepared by Richard Schmitt and recently published as "Frege's Letters to Wittgenstein about the *Tractatus*," *Bertrand Russell Society Quarterly* 120 (Nov. 2003): 13–31.

List of the Extant Letters and Postcards from Frege to Wittgenstein

11 October 1914	Jena	Army Postcard
23 December 1914	Jena	Army Postcard
24 June 1915	Jena	Army Postcard
28 November 1915	Jena	Army Postcard
6 February 1916	Jena	Army Postcard
21 April 1916	Brunshaupten	Army Postcard
2 July 1916	Jena	Army Postcard
29 July 1916	Jena	Army Postcard
28 August 1916	Jena	Army Postcard
26 April 1917	Brunshaupten	Army Postcard
30 June 1917	Brunshaupten	Letter
16 September 1917	Brunshaupten	Army Postcard
26 February 1918	Neuburg, Mecklenburg	Army Postcard
9 April 1918	Neuburg, Mecklenburg	Letter
1 June 1918	Neuburg, Mecklenburg	Army Postcard
12 September 1918	Neuburg, Mecklenburg	Postcard
15 October 1918	Bad Kleinen	Army Postcard
28 June 1919	Bad Kleinen	Letter
16 September 1919	Bad Kleinen	Letter
30 September 1919	Bad Kleinen	Letter
3 April 1920	Bad Kleinen	Letter

Chronology of the Known Frege-Wittgenstein Correspondence

Boldface	signifies an extant piece of correspondence, now housed either at the Brenner archives in Innsbruck or the Scholz archives in Münster and translated below
"⋆"	signifies a piece of correspondence assumed no longer to exist
"[ ]"	signifies comments made by the collector of Frege's papers, Heinrich Scholz, concerning the contents of the piece of correspondence
Italics	signify a relevant passage in an independent source making reference to the correspondence with Frege
"{}"	signifies sources of evidence for the conjectured existence and contents of pieces of correspondence that are now lost. For further detail readers are encouraged to consult the editorial comments given in the Intelex electronic version of Wittgenstein's correspondence, as well as editorial remarks by G. Gabriel in his previous presentation of the correspondence in G. Gabriel et al., eds., *Gottlob Frege, Wissenschaftlicher Briefwechsel* vol. II of *Nachgelassene Schriften und Wissenschaftlicher Briefwechsel* (Hamburg: Felix Meiner Verlag, 1976), hereafter "WB". (The roman numeral "XLV" followed by an Arabic numeral refers to Gabriel's numbering of the Frege-Wittgenstein correspondence in this edition.) Quotations from letters of Wittgenstein to Bertrand Russell may be found also in B. McGuinness and G. H. von Wright, eds., *Ludwig Wittgenstein, Cambridge Letters: Correspondence with Russell, Keynes, Moore, Ramsey and Sraffa* (Cambridge, MA: Blackwell, 1995).

*1911(?) Wittgenstein to Frege

{"I wrote to Frege, putting forward some objections to his theories, and waited anxiously for a reply...", comment reported by Peter Geach; cf. Peter Geach, "Frege", in G.E.M. Anscombe and P.T. Geach, *3 Philosophers* (Oxford: Basil Blackwell, 1961), pp. 127–162; passage is from pp. 129–130.}

*1911(?) Frege to Wittgenstein

{"To my great pleasure, Frege wrote and asked me to come and see him ... I was shown into Frege's study. Frege was a small neat man with a pointed beard, who bounced around the room as he talked. He absolutely wiped the floor with me, and I felt very depressed; but at the end he said 'You must come again', so I cheered up. I had several discussions with him after that." (cf. Peter Geach, "Frege" (op. cit.), pp. 129–130)}

*22 October 1913 Wittgenstein to Frege Skjolden

{XLV/1}

["Concerning a request for a visit."]

*??? ["4 pages of notes by Frege on the Wittgensteinian standpoint that was communicated orally"] {Wittgenstein to Bertrand Russell 26.12.12: "I had a long discussion with Frege about our Theory of Symbolism of which, I think, he roughly understood the general outline. He said he would think the matter over. The complex problem is now clearer to me and I hope very much that I may solve it." Compare McGuinness, *Wittgenstein, A Life: Young Ludwig 1889–1921* (Berkeley, CA: University of California Press, 1988), p. 164: "It seems very probable that it was on this occasion that Frege put to Wittgenstein a general objection to Russell's talk about complexes, to which Wittgenstein reverted many times in conversation and alluded also in his notebooks. Frege asked him whether, if an object is part of a fact about it, the fact will be bigger than the object. At the time he thought the remark silly, but later he came to see the point of it. It was in fact an attack on the whole notion of explaining the meaning of propositions by saying that there were complexes corresponding to them—a way of speaking that Wittgenstein did in fact abandon in the course of 1913."}

*9 November 1913 Frege to Wittgenstein

{XLV/2}

["Fragment of a letter, probably from 9.11.13. Contents: Reply to 22 October 1913. Reproof, W. lays too great value upon signs."]

*29 November 1913 Wittgenstein to Frege Skjolden
{XLV/3}
["Contents: *Important* arguments against Frege's theory of truth. Especially against the determination of meaning [*Bedeutungsfestsetzung*] for functions."]

*beginning of
December 1913 Wittgenstein to Frege Skjolden
{XLV/4}
["Contents: Reply to {9 November 1913}. On the drawing up of a set of fundamental concepts of logic and the requirements that are to be imposed on them. Announcement of a visit."] {This visit, in December 1913, was presumably the last time that Frege and Wittgenstein met. Cf. Geach, "Frege", op. cit.: "The last time I saw Frege, as we were waiting at the station for my train, I said to him 'Don't you ever find any difficulty in your theory that numbers are objects?' He replied 'Sometimes I *seem* to see a difficulty—but then again I *don't* see it.' "}

*end of Jan.
or early Feb. 1914 Frege to Wittgenstein Brunshaupten
{XLV/5}
["4 pages beginning a letter. Contents: continuation of oral conversation."] {Gabriel, ed., WB, p. 266n: "it remains unclear whether the four pages belonged to a draft and the letter of Frege's was ever finished and sent."}

*before
11 October 1914 Wittgenstein to Frege Kraków
{Army postcard, mentioned in Frege to Wittgenstein, 11.10.1914}

11 October 1914 Frege to Wittgenstein Jena
{Cf. Wittgenstein, *Geheime Tagebücher 1914–1916*, ed. W. Baum (Vienna: Turia & Kant Verlag, 1992) p. 37 (in von Wright's numbering of Wittgenstein's papers this is MS 102), an entry from 30.10.14: "(In the evening) got some very dear mail, a very dear card from Frege! One from Trakl and Ficker, Mama, Clara, Mrs. Kingenberg. This made me very happy. Worked very hard."}

*before
23 December 1914 Wittgenstein to Frege Kraków
{Army postcards, the number unknown; mentioned in Frege to Wittgenstein of 23.12.1914}

23 December 1914 Frege to Wittgenstein Jena

*before
24 June 1915 Wittgenstein to Frege
{Army postcards, the number unknown; mentioned in Frege to Wittgenstein of 24.6.1915}

24 June 1915 Frege to Wittgenstein Jena

*25 August 1915 Wittgenstein to Frege
{XLV/6}
["Army postcard from 25.8.1915."]
["Wittgenstein to Frege on his treatise. New address."]
{Gabriel, ed., WB, p. 266: "this army postcard begins Wittgenstein's correspondence with Frege about Wittgenstein's "*Abhandlung*", later known as the "*Tractatus*". This is shown in Scholz's way of organizing his hand-written description of this series of cards. B. McGuinness and J. Schulte, eds., *Logisch-philosophische Abhandlung/Tractatus Logico-Philosophicus, Kritische Edition* (Frankfurt, Suhrkamp Verlag, 1989) p. xxii: "all the evidence points toward the idea that the '*Abhandlung*' on which Wittgenstein was working in October 1915 was preserved and presented in the first 70 pages of the *Prototractatus*, which was probably written at this time." That is, between the fall of 1915 and March of 1916 when Wittgenstein departed for the front}

*before
28 November 1915 Wittgenstein to Frege
{Army postcards, the number unknown; mentioned in Frege to Wittgenstein, 28.11.1915}

28 November 1915 Frege to Wittgenstein Jena

*before
6 February 1916 Wittgenstein to Frege
{Army postcard, mentioned in Frege to Wittgenstein, 6.2.1916}

6 February 1916 Frege to Wittgenstein Jena

*before
21 April 1916 Wittgenstein to Frege
{Letter, mentioned in Frege to Wittgenstein, 21.4.1916}

*before
21 April 1916 Wittgenstein to Frege

	{Army postcard, mentioned in Frege to Wittgenstein, 21.4.1916}	
21 April 1916	**Frege to Wittgenstein**	**Brunshaupten**
*before 2 July 1916	Wittgenstein to Frege {Army postcards, the number unknown; mentioned in Frege to Wittgenstein, 2.7.1916}	
2 July 1916	**Frege to Wittgenstein**	**Jena**
*before 29 July 1916	Wittgenstein to Frege {Army postcard, mentioned in Frege to Wittgenstein, 29.7.1916}	
29 July 1916	**Frege to Wittgenstein**	**Jena**
*16 August 1916	Wittgenstein to Frege {Army postcard, mentioned in Frege to Wittgenstein, 28.8.1916}	
28 August 1916	**Frege to Wittgenstein**	**Jena**
*before 26 April 1917	Wittgenstein to Frege {unknown number of army postcards, mentioned in Frege to Wittgenstein, 26.4.1917}	
26 April 1917	**Frege to Wittgenstein**	**Brunshaupten**
*15 June 1917	Wittgenstein to Frege {Letter, mentioned in Frege to Wittgenstein, 30.6.1917}	
30 June 1917	**Frege to Wittgenstein** {Letter; two half-pages are missing; contained a picture}	**Brunshaupten**
*5 September 1917	Wittgenstein to Frege {Army postcard, mentioned in Frege to Wittgenstein, 16.9.1917}	
16 Sept. 1917	**Frege to Wittgenstein**	**Brunshaupten**

*8 February 1918	Wittgenstein to Frege {Army postcard, mentioned in Frege to Wittgenstein, 26.2.1918}	
26 February 1918	**Frege to Wittgenstein**	**Neuburg bei Wismar**
*25 March 1918	Wittgenstein to Frege {Letter, mentioned in Frege to Wittgenstein, 9.4.1918}	
9 April 1918	**Frege to Wittgenstein**	**Neuburg bei Wismar**
*10 May 1918	Wittgenstein to Frege {Army postcard, mentioned in Frege to Wittgenstein, 1.6.1918}	
1 June 1918	**Frege to Wittgenstein**	**Neuburg bei Wismar**
*before 12 September 1918	Frege to Wittgenstein {Army postcard, mentioned in Frege to Wittgenstein, 12.9.1918}	
*before 12 September 1918	Wittgenstein to Frege {Letter, mentioned in Frege to Wittgenstein, 12.9.1918}	Bozen
12 Sept. 1918	**Frege to Wittgenstein** {XLV/7}	**Neuburg (Mecklenburg)**
*between 12 September and 12 October 1918	Wittgenstein to Frege {XLV/8} ["Army postcard, Wittgenstein to Frege". "Thanks for card of 12.9.1918"]	Vienna {?}
*12 October 1918	Wittgenstein to Frege {XLV/9} ["Army postcard, Wittgenstein to Frege"]	
15 October 1918	**Frege to Wittgenstein**	**Bad Kleinen**
*26 October 1918	Wittgenstein to Frege	

{XLV/10}
["Army postcard, Wittgenstein to Frege from 26.10.1918"]

{XLV/11 is a letter from Wittgenstein's sister Hermine to Frege of 24.12.1918, communicating that Wittgenstein was placed in a prisoner of war camp and that a typescript of his "work" (later known as the *Tractatus*) would be sent to Frege. Cf. Gabriel, ed., WB, p. 266.}

*23 February 1919 Wittgenstein to Frege Monte Cassino
{XLV/12}
["Card, Wittgenstein to Frege from 23.2.1919"]

{XLV/13 is a letter from Wittgenstein's sister Hermine to Frege of 19.3.1919.}
{XLV/14 is a card from Hermine to Frege. ["the missing pages of the work have been sent"]}

*10 April 1919 Wittgenstein to Frege Monte Cassino
{XLV/15}
["Card from Wittgenstein to Frege. Request for judgment about the work."]

*9 June 1919 Wittgenstein to Frege Monte Cassino
{XLV/16}
["Card from Wittgenstein to Frege."]

28 June 1919 Frege to Wittgenstein Bad Kleinen

{XLV/17 is a letter from Wittgenstein's sister Hermine to Frege of 7.7.1919}
{XLV/18 is a letter from Wittgenstein's sister Hermine to Frege of 17.7.1919; ["The letter of Frege's to Wittgenstein was sent along."]}

*3 August 1919 Wittgenstein to Frege Monte Cassino
{XLV/19}
["Answer to the letter of Frege's sent by the sister."]

{XLV/20 is a letter from Wittgenstein's sister Hermine to Frege of 28.8.1919; ["Wittgenstein returned from prison."]}

*6 September 1919 Wittgenstein to Frege Vienna
{XLV/21}
["Letter, Wittgenstein to Frege. About his treatise."]

16 Sept. 1919 **Frege to Wittgenstein** **Bad Kleinen**

*16 September 1919 Wittgenstein to Frege Vienna
{XLV/22}
["Letter, Wittgenstein to Frege of 16.9.1919. Thanks for 'Der Gedanke', critical remarks on it. Request to apply to publish his treatise in the BPhDI"] {That is, in the *Beiträge zur Philosophie des deutschen Idealismus.*}
{Gabriel, ed., WB, p. 268n: See the letter III/4 (Bauch to Frege 31.10.1919) and XVI/3 (Hoffmann to Frege 23.1.1920).}

30 Sept. 1919 **Frege to Wittgenstein** **Bad Kleinen**

*between 13 and
20 December 1919 Wittgenstein to Frege The Hague
{XLV/23}
["Letter from Wittgenstein to Frege. Time of the meeting with Russell. With the signature of Russell. Announcement of a visit from Wittgenstein."]
{G.H. von Wright, "The Origin of the *Tractatus*", in *Wittgenstein* (Minneapolis, MN: University of Minnesota Press, 1983), pp. 63–110, cf. p. 88: "From the meeting with Russell in The Hague Wittgenstein also wrote to Frege, announcing a visit on the way home to Austria. The visit, however, had to be cancelled, because Wittgenstein's companion on the journey to Holland, Mr. Arvid Sjögren, had fallen seriously ill. Wittgenstein arrived back from the meeting with Russell on 26 or 27 December."}

*29 December 1919 Wittgenstein to Frege Vienna
{XLV/24}
["Letter from Wittgenstein to Frege. Report on the meeting with Russell, who will possibly get his book published in England."]

*19 March 1920 Wittgenstein to Frege Vienna
[Letter, mentioned in Frege to Wittgenstein, 3.4.1920]

3 April 1920 **Frege to Wittgenstein** **Bad Kleinen**

2 April 1936 **Scholz to Wittgenstein** **Münster**

9 April 1936 **Wittgenstein to Scholz** **Cambridge**

{It is difficult to know how to date the conversation alluded to by Wittgenstein in *Philosophical Remarks* XI (Oxford: Basil Blackwell, 1975), but his remark (penned in 1930) is this: "... the Fregean theory of number would be applicable provided we were not intending to give an analysis of propositions. This theory explains the concept of number for the idioms of everyday speech. Of course, Frege would have said (I remember a conversation we had) that the simultaneous occurrence of an eclipse of the moon and a court case was an object. And what's wrong with that? Only that we in that case use the word 'object' ambiguously, and so throw the results of the analysis into disarray."}

Frege-Wittgenstein Correspondence

translated by *Burton Dreben and Juliet Floyd*

E. De Pellegrin (ed.), *Interactive Wittgenstein*, Synthese Library 349,
DOI 10.1007/978-1-4020-9909-0_2,

Prof. Frege
Jena, Forstweg Nr. 29

An den Kriegsfreiwilligen
Ludwig Wittgenstein
Militär Kommando Krakau
Feldpost Nro 186

Jena, d. 11.X.14

Lieber Herr Wittgenstein! Ich danke Ihnen bestens für Ihren Kartengruss. Dass Sie als Kriegsfreiwilliger eingetreten sind, habe ich mit besonderer Befriedigung gelesen und bewundere es, dass Sie sich noch dabei der Wissenschaft widmen können. Möge es mir vergönnt sein, Sie nach dem Kriege gesund wiederzusehen, und die Unterredungen mit Ihnen weiterzuführen. Gewiss werden wir uns dadurch zuletzt näher kommen und uns immer besser verstehen. Wir hatten hier 3 Leichtverwundete im Hause; Alfred musste dazu seine Spielstube hergeben. Sie erzählten viel von ihren Kämpfen in den Vogesen, ohne Hochachtung vor den Franzosen, denen sie sich bei gleicher Anzahl überlegen fühlten; aber sie hatten den Eindruck, meist gegen eine grosse Überzahl gekämpft zu haben. Ihnen alles Gute wünschend mit herzlichem Grusse

Ihr G. Frege

Army Postcard

Prof. Frege
Jena, Forstweg 29

To the Volunteer
Ludwig Wittgenstein
Kraków Military Command
Field Post Number 186

Jena, 11 October 1914

Dear Mr. Wittgenstein: Thank you very much for your greetings by card. I read with special satisfaction that you enlisted in the war as a volunteer, and I marvel that you are still able to pursue science. I hope I shall have the privilege of seeing you again after the war, in good health, and of pursuing our conversations further. Surely in this way we will eventually become closer and come to understand one another better and better. We had three slightly wounded people at our home; hence Alfred had to give up his playroom. They told us much about their battles in the Vosges, showing no respect for the French to whom they felt superior when matched in equal numbers; however, they had the impression of being in battle for the most part against vastly greater numbers. Wishing you all the best with kind regards,

Yours, G. Frege

Professor Dr. G. Frege
Jena, Forstweg Nr. 29

An Ludwig Wittgenstein
K. u. K. Werkstätte der Festung Krakau
Artillerie-Auto-Detachement

Jena, d. 23.XII.14

Lieber Herr Wittgenstein! Herzlichen Dank für Ihre Karten! In diesem Jahre ist die Weihnachtsstimmung nicht so fröhlich wie sonst. Von meinen Verwandten is[t] einer als Pionier bei der Eroberung Antwerpens gefallen, ein anderer ist in Polen verwundet und liegt in Schlesien. Einer steht im Felde, sein Bruder trägt gut 8 Wochen den bunten Rock und wird ihn bald mit dem feldgrauen vertauschen. In der Wismarschen Zeitung las ich zu Anfang des Krieges folgendes Gespräch eines Reservemannes, der den Eisenbahnzug bestiegen hatte mit seiner Frau. Es war plattdeutsch geführt; ich gebe es hochdeutsch wieder. Mann: "Sieh zu, dass du den Roggen gut herein kriegst." Frau: "Geht dich garnichts an. Pass du nur auf, dass du die Patronen nicht verschwendest, dass jede Kugel trifft."—Es freut mich, dass Sie in dieser schweren Zeit immer noch Zeit und Kraft zur wissenschaftlichen Arbeit haben; mir will es nicht recht gelingen.—Wünschen wir uns den Sieg unserer Krieger und einen dauerhaften Frieden im nächsten Jahre.

Beste Grüsse sendet Ihnen

Ihr G. Frege.

Army Postcard

Professor Dr. G. Frege
Jena, Forstweg 29

To Ludwig Wittgenstein
K. u. K. Workshop of the Kraków Fortress
Artillery-Auto-Detachment

Jena, 23 December 1914

Dear Mr. Wittgenstein: Many thanks for your cards! This year the Christmas spirit is not as joyful as usual. As for my relatives, one fell as an engineer in the taking of Antwerp, another, wounded in Poland, is in the hospital in Silesia. Yet another is in the field; his brother has been in military service for 8 weeks and will soon be wearing the field grey. At the beginning of the war in the *Wismarsche Zeitung* I read of the following conversation a reservist had with his wife as he was boarding the train. It was recounted in Low German; I shall tell it in High German. Husband: "See to it that you bring in the rye." Wife: "That's not your worry. You just take care that you don't waste rounds, that every bullet hits."—I am pleased that even in these difficult times you still have time and energy for scientific work. I do not really.—For next year let us hope for a victory of our warriors and a lasting peace.

Best wishes to you,

Yours, G. Frege

Professor Frege
Jena, Forstweg 29

An Herrn L. Wittgenstein
K. u. K. Art. Werkstätte der Festung Krakau

Jena, d. 24.VI.15

Sehr geehrter Herr Wittgenstein!

Herzlichen Dank für Ihre Karten! Es freut mich, dass Sie wissenschaftlich arbeiten. Ich kann das Gleiche von mir kaum sagen. Ich weiss nicht, wie es zugeht, dass ich zu nichts recht Zeit habe. Ich habe hier eine Menge Feldpostkarten liegen, die [ich] beantworten müsste, und komme nicht dazu. Meist sind sie von Verwundeten, die hier im Hause waren. Einer von diesen ist nachher zum zweiten Male verwundet worden, ein Anderer ist gefallen (in Polen). Über die Wiedereroberung Lembergs haben wir uns sehr gefreut. Gegenüber all dem Heldenmute der jetzt betätigt wird, kommt mir mein Tun nichtig und unbefriedigend vor. Mit Ihnen noch einmal in friedlichen Zeiten wissenschaftliche Gespräche führen zu können, hofft

Ihr G. Frege

Army Postcard

Professor Frege
Jena, Forstweg 29

To Mr. L. Wittgenstein
K u. K. Art. Workshop of the Kraków Fortress

Jena, 24.VI.15[1]

Dear Mr. Wittgenstein,

Many thanks for your cards! I am pleased that you are doing scientific work. I can hardly say the same for myself. I do not know how it happens that I never have enough time for anything. I have lying here quite a collection of army postcards that I ought to answer, and do not get to. Mostly they are from wounded who were here in the house. One of these was subsequently wounded for the second time, another fell in combat (in Poland). We were very pleased about the recapture of Lemberg.[1] Compared with all the heroism now being exercised, my activities seem to me pointless and unsatisfying. To be able to carry on scientific conversations with you again in peaceful times is the hope of

Your G. Frege

[1] The card's official postmark says "25.6.16", but this is presumed to be an error. We take the date to be what Frege writes, viz., "24.VI.15". Reasons for preferring the earlier date are as follows. First, Wittgenstein was not in the Kraków workshop after July 1915 (on this see Brian McGuinness, *Wittgenstein, A Life: Young Ludwig 1889–1921* (Berkeley, CA: University of California Press, 1988) and Ray Monk, *Wittgenstein: The Duty of Genius* (New York: The Free Press, 1990), p. 132). Second, Lemberg was recaptured in 1915, and not 1916, and it seems most likely that Frege is referring in the letter to this recent event. ("Lemberg", now in the Ukrainian Democratic Republic, is currently known as "Lviv"; when it was part of Poland (1919–1945) it was known as "Lvov".)

An Herrn Wittgenstein
K. u. K. Art. Werkstätten Zug Nro 1
Feldpost Nro 12 Österreich

Jena, d. 28. Nov. 1915.

Sehr geehrter Herr Wittgenstein!

Vielen Dank für Ihre Feldpost-Karten. Es freut mich, dass Sie immer noch Zeit und Kraft für wissenschaftliche Arbeiten übrig haben. Zugleich ist mir dies ein Zeichen für Ihr Wohlergehen. Auch mir geht es—abgesehen von Kleinigkeiten—gut. Hoffentlich werden wir uns nach dem Kriege in guter Gesundheit wiedersehen und Sie mir dann viel von Ihren äusseren und inneren Taten berichten können. Was ich dagegen zu bieten haben werde, wird freilich nur gering sein. Mit besten Wünschen für Ihr ferneres Wohlergehen grüsst Sie herzlich

Ihr G. Frege

Army Postcard

To Mr. Wittgenstein
K. u. K. Art. Workshop Train Number 1
Field Post Number 12, Austria

Jena, 28 November 1915

Dear Mr. Wittgenstein,

Many thanks for your army postcards. I am pleased that you still have time and energy for scientific work. This strikes me as a sign of your well-being. I am also well—apart from trivial matters. Let us hope that we shall see each other again after the war, in good health, and that you will then be able to tell me much about your outer and inner deeds. What I shall have to offer in exchange will of course be little. With best wishes for your continued well-being and kind regards,

Yours, G. Frege

Frege, Jena, Forstweg Nr 29

An Ludwig Wittgenstein
K. u. K. A. W. Z. Nro. 1
Feldpost Nro 12

Jena, d. 6. Febr. 1916

Sehr geehrter Herr Wittgenstein!

Verzeihen Sie, dass ich erst jetzt dazu komme, Ihre Karte zu beantworten. Ich habe mich mit diesen Sachen auch nicht beschäftigt. Vielleicht handelt es sich um Folgendes. Man hat unendl. viele Gruppen von 3 Zahlen a, b, c, sodass $a^2 + b^2 = c^2$. Es handelt sich nun vielleicht darum zu beweisen, dass eine solche Gleichung wie $a^n + b^n = c^n$ durch ganze Zahlen nicht zu erfüllen ist, wenn $n > 2$ ist.

Herzlich grüsst Sie

Ihr G. Frege

Army Postcard

Frege, Jena, Forstweg No. 29

To Ludwig Wittgenstein
K. u. K. A. W. T. Number 1
Field Post Number 12

Jena, 6 February 1916

Dear Mr. Wittgenstein,

Forgive me for not replying to your card until now. I have not worked on these matters. Perhaps it has to do with the following. There are infinitely many groups of three numbers a, b, c such that $a^2 + b^2 = c^2$. It would perhaps come down to proving that an equation such as $a^n + b^n = c^n$ cannot be satisfied by any whole number if $n > 2$.

Kind regards to you

Yours, G. Frege

Prof. Dr. Frege, Jena

An Herrn Ludwig Wittgenstein
K. u. K. Feldkanonenregiment Nro 2
Btt 4, Feldpost Nro 72

Brunshaupten, den 21.IV.16

Sehr geehrter Herr Wittgenstein!

Besten Dank für Ihren Brief u. Ihre Karte. Ihren Wunsch, Ihre geistige Arbeit nicht verlorengehen zu lassen, finde ich sehr begreiflich und möchte gern das Meinige dazu beitragen. Aber ob ich nach Wien kommen kann, ist mir noch zweifelhaft. Vielen Dank für Ihre freundl. Einladung. Jedenfalls hoffe ich, dass es mir in irgendeiner Weise vergönnt sein möge, unsere wissenschaftlichen Unterhaltungen weiter zu führen, und dann müssen wir uns ja auch mit der Zeit näher kommen. Demnächst komme ich wieder nach Jena.

Mit herzlichem Grusse

Ihr G. Frege

Army Postcard

Prof. Dr. Frege, Jena

To Mr. Ludwig Wittgenstein
K. u. K. Field Artillery Regiment Number 2
Btt 4, Field Post Number 72

Brunshaupten, 21.IV.16

Dear Mr. Wittgenstein,

Many thanks for your letter and your card. I find your hope not to let your intellectual work be lost very understandable, and I would very much like to contribute what help I can. However, I still doubt that I can come to Vienna. Many thanks for your friendly invitation. In any case I hope that in some way or other I shall have the privilege of further pursuing our scientific conversations, and then in time we are bound to become closer. I shall return to Jena soon.

With kind regards,

Yours, G. Frege

Forstweg Nro 29

An Ludwig Wittgenstein
K. u. K. Feldhaubitzen Regiment Nr. 5 Batterie Nro 4
F. H. R. 5/4 Feldpost Nro 110

Jena 2.VII.16

Lieber Herr W! Besten Dank für Ihre Karten! Leider fehlt darin die frühere gute Stimmung. Ich hoffe sehr, dass Sie sie bald wiedergewinnen im erfolgreichen Kampfe für eine grosse Sache in einer weltgeschichtlichen Entscheidung, wie es noch keine gegeben hat. Auch ich habe jetzt nicht recht Kraft und Stimmung zu eigentlich wissenschaftlichen Arbeiten, suche mich aber zu betätigen in der Ausarbeitung eines Planes, von dem ich hoffe, dass er dem Vaterlande nach dem Kriege nützlich sein kann. Dann hoffe ich, dass wir unsere Gespräche zur gegenseitigen Verständigung und Förderung in logischen Fragen wieder aufnehmen können.

Mit den besten Wünschen für Ihr Wohlergehen

Ihr G. Frege

Army Postcard

Forstweg No. 29

To Ludwig Wittgenstein
K. u. K. Field-howitzer Regiment Number 5 Battery Number 4
F. H. R. 5/4 Field Post Number 110

Jena 2.VII.16

Dear Mr. W! Many thanks for your cards! I am sorry that your earlier high spirits are missing from them. I very much hope that you regain these soon in the successful struggle for a great cause in a decisive world-historical context the likes of which there has never been. Right now I too lack enough strength and frame of mind for genuinely scientific work, but I am trying to occupy myself by working out a plan that I hope may be useful to the Fatherland after the war.[2] Then I hope that we shall be able to resume our conversations so as to make progress on our mutual understanding and on logical questions.

With best wishes for your well-being,

Yours, G. Frege

[2] Nothing in particular is known about this plan. Frege did record various political thoughts in his diary. These were published with annotated commentary in "Gottlob Freges politisches Tagebuch. Mit Einleitung und Kommentar", eds. G. Gabriel and W. Kienzler, *Deutsche Zeitschrift für Philosophie* 42 (1994), 6: 1057–1098, an edition which subsequently appeared in an English translation by Richard L. Mendelsohn in *Inquiry* 39 (1996): 303–342. For more on Frege's politics in historical context, see Wolfgang Kienzler, "Frege und Deutschland" in K.M. Kodalle, ed., *Die Angst vor der Moderne. Philosophische Antworten auf Krisenerfahrungen. Der Mikrokosmos Jena* 1900–1940 (Königshausen & Neumann, Würzburg, 2000), pp. 135–156. On Frege's liberalism before 1918 one may read his proposal for an election system in "Vorschläge für ein Wahlgesetz von Gottlob Frege", eds. U. Dathe and W. Kienzler, in G. Gabriel and U. Dathe, eds., *Gottlob Frege-Werk und Wirkung* (Paderborn: Mentis Verlag, 2000), pp. 283–313. Compare Uwe Dathe, "Wismar, Jena, Bad Kleinen—wo liegen die Wurzeln für Gottlob Freges politische Anschauungen? Einige Ergänzungen zu Lothar Kreisers Frege-Biographie", *Zeitschrift des Vereins für Thüringische Geschichte* 56 (2002): 417–421, a comment which should be compared with Lothar Kreiser's discussion in *Gottlob Frege Leben-Werk-Zeit* (Hamburg: Felix Meiner Verlag, 2001).

An L. Wittgenstein
K. u. K. Feldhaubitzen-Regiment Nr. 5 4 Batterie
Feldpost 47

Jena, den 29. Juli 16

Lieber Herr Wittgenstein! Ich danke Ihnen bestens für Ihre Grüsse. Ich freue mich immer, wenn ich ein Lebenszeichen von Ihnen erhalte. Verzeihen Sie mir nur, dass meine Antworten so selten sind. Mir geht in dieser Zeit immer so vielerlei durch den Kopf, dass ich zum Kartenschreiben nur zu selten komme, obwohl äusserlich das Leben seinen gewohnten Gang geht. Hoffentlich erhalte ich bald einmal von Ihnen eine Karte, aus der ich eine recht gehobene Stimmung herauslese.

Mit bestem Grusse

Ihr G. Frege

Jena, Forstweg Nro 29

Army Postcard

To Ludwig Wittgenstein
K. u. K. Field-howitzer Regiment Number 5 4 Battery
Field Post 47

Jena, 29 July 16

Dear Mr. Wittgenstein:

I thank you very much for your greetings. I am always pleased when I get a sign of life from you. But do forgive me that I reply to you so infrequently. Although on the surface life goes on as usual, so much runs through my head right now that I rarely get around to writing cards. I hope I shall soon receive another card from you in which I read of your truly high spirits.

With best regards,

Yours, G. Frege

Jena, Forstweg Number 29

An Herrn Ludwig Wittgenstein
K. u. K. Feldhaubitzen Regiment Nr. 5 4 Batterie
Feldpost Nro 72
Olmütz 4

Jena, d. 28.VIII.16.

Sehr geehrter Herr Wittgenstein!

Besten Dank für Ihre Karte vom 16. d.! Soeben höre ich, dass Rumänien den Krieg an Österreich erklärt hat. Nun gilt es, alle Kraft zusammenzunehmen! und den Nacken steif halten! Dies regt mich so auf, dass zunächst kaum etwas anderes Raum in mir hat. Können Sie nicht dann und wann soviel Zeit erübrigen, dass Sie Ihre Gedanken, wenn auch abgerissen und ungeordnet zu Papier bringen und mir mitteilen können? Ich werde dann diese Briefe für Sie aufbewahren und versuchen, Ihnen zu antworten. So würde doch vielleicht ein wissenschaftlicher Verkehr zwischen uns herzustellen sein und so wenigstens ein geringer Ersatz für die mündliche Aussprache geschaffen werden.

Mit bestem Grusse Ihr

G. Frege.

Army Postcard

To Mr. Ludwig Wittgenstein
K. u. K. Field-howitzer Regiment Number 5 4 Battery
Field Post Number 72
Olmütz 4

Jena, 28.VIII.16

Dear Mr. Wittgenstein,

Many thanks for your card of the 16th! I have just heard that Romania has declared war on Austria. Now we must summon all our strength and stiffen our backs! This so disturbs me that at present I can think of little else. Could you not now and then spare some time to set down your thoughts—even if disjointed and unordered—on paper, and communicate them to me? I will preserve these letters for you and attempt to answer you. In this way a scientific exchange might be brought about between us that would be at least a small substitute for face-to-face conversation.

With best regards,

Yours, G. Frege

Prof. Dr. Frege aus Jena
z. Z. in Brunshaupten (Mecklenb.)
Villa Vineta

An Wittgenstein Fhrch
F. H. R. 5/4
Feldpost N° 286
K. u. K. Husaren-Rgt Nr 16?
Schützeneskadr.
Österreichisches Heer

Brunshaupten, d. 26.IV.17
Villa Vineta

Lieber Herr Wittgenstein! Besten Dank für Ihre Karten. Ich bewundere Ihre Wandlungsfähigkeit: in der Festung Krakau, auf der Weichsel mit Scheinwerfern, bei den Feldkanonen, bei den Feldhaubitzen und nun bei den Husaren. Und dabei finden Sie noch Zeit zu wissenschaftlichen Arbeiten! Das scheint Ihnen besser zu gelingen als mir. Die Runen auf Ihrer Karte, die Ihre Anschrift angeben, sind schwer zu enträtseln. Hoffentlich gelangt diese Karte trotz aller Hindernisse in Ihre Hände. Ob der lange Krieg sich nun endlich dem Ende zuneigt? Einiges spricht dafür, die Auflösung in Russland, die Erfolge des U-bootkrieges, der misslungene Durchbruch der Engländer u. Franzosen im Westen. Hoffen wir das Beste! Und dazu wird auch gehören, dass wir unseren Gedankenaustausch wieder aufnehmen können.

Mit besten Grüssen

Ihr G. Frege

Army Postcard

Prof. Dr. Frege from Jena
presently in Brunshaupten (Mecklenb.)
Villa Vineta

To Wittgenstein Fhrch[3]
F. H. R. 5/4
Field Post Number 286
K. u. K. Hussar-Regiment Number 16?
Schützeneskadr.
Austrian Army

Brunshaupten, 26.IV.17
Villa Vineta

Dear Mr. Wittgenstein! Many thanks for your cards. I admire your capacity for change: in the Kraków fortress, on the Weichsel[4] with searchlights, with the field cannons, with the field howitzers and now with the Hussars. And yet you still find time for scientific work! It does seem that you are more successful at that than I am. The runes on your cards which give your address are difficult to decipher. Despite all difficulties I hope this card will successfully reach you. Will the long war now finally draw to a close? Some things speak for this: the dissolution in Russia, the success of the U-boat war, the failed breakthrough of the English and French in the West. Let us hope for the best! Part of that will be that we shall be able to resume our exchange of thoughts.

With best regards,

Yours, G. Frege

[3] "Fhrch" probably an abbreviation for "Fähnrich". Wittgenstein was given the rank of Fähnrich in the Reserve on 1 December 1916 (later backdated to 1 October 1916) according to McGuinness, p. 256.

[4] That is, on the river Vistula.

Brunshaupten (Ostsee), den 30.VI.17
(Villa Anna-Lise)

Lieber Herr Wittgenstein!

Erst gestern habe ich Ihren Brief vom 15.VI erhalten und nun ist schon die Hälfte Ihrer Urlaubszeit verflossen. Sehr schwer wird es mir, Ihrer liebenswürdigen Einladung nicht zu folgen, aber noch schwerer, ihr zu folgen. Ich fühle mich angegriffen und bin hier zu meiner Erholung, die ich nötig habe. Die Reise nach Wien und wieder zurück ist mir unter diesen Umständen zu angreifend. Vielleicht würden Sie auch von meinem geistigen [...]

mein leiblicher und geistiger Zustand dazu geeignet sein mag. Ich habe für das Sommersemester Urlaub und sehe dem Winter in Jena nicht sehr zuversichtlich entgegen.

Mit besten Wünschen für Ihr Wohlergehen und Dank für das hübsche Bildchen, das ich neulich von Ihnen erhielt, hofft auf ein fröhliches Wiedersehn

Ihr ergebener G. Frege.

[...]

Ihre militärische Dienst-Anschrift zu benutzen. Da ich aber nicht weiss, ob die letzte mir bekannte noch gilt, schicke ich den Brief wieder nach Wien, wie Sie mir angegeben haben. Hoffentlich wird er Sie dort noch erreichen.

Mit besten Grüssen

Ihr G. Frege

Letter[5]

Brunshaupten (Ostsee), 30.VI.17
(Villa Anna-Lise)

Dear Mr. Wittgenstein,

I received your letter of 15.VI only yesterday, and now half of your leave time has already gone by. It is very hard for me not to accept your gracious invitation, but still harder to accept it. I feel exhausted and am here for a necessary recuperation. The journey to Vienna and back again would in these circumstances be too exhausting for me. You would also possibly [...] my mental

[...]

to which my physical and mental condition may be better suited. I am on leave in the summer semester and do not face the winter in Jena very confidently.

With best wishes for your well-being and thanks for the small pretty picture that I recently received from you. Hoping for a cheerful reunion soon,

Devotedly yours, G. Frege

[...]

to use your military service address. Because I do not know whether the last one I knew is still valid, I shall send the letter again to Vienna as you instructed me. I hope it will still reach you there.

With best regards,

Yours, G. Frege

[5] This letter appears to have consisted of one bifolium page. On the first side only the right sector was written on; on the second, both sectors were written on and both were signed by Frege; presumably the second signature attached to a postscript. Only the top part of the letter remains: a portion of the bottom of the page is torn off, quite neatly, along a line. I have put square brackets with an ellipsis mark at the points in the text of the letter that are missing. It is possible that this was done intentionally in an effort to blot from the record a detailed description by Frege of his mental condition, which was not always stable (see L. Kreiser, *Gottlob Frege Leben-Werk-Zeit*, pp. 512ff). The editors of the CD-ROM of Wittgenstein's *Briefwechsel* report that this letter "contained a picture."

Prof. Frege aus Jena
Brunshaupten in Mecklenb
Neue Reihe Nr 208

An Wittgenstein Fhrch
F. H. R. 5/4
Feldpost 286
Österreich-Ungarn

Brunshaupten, den 16.IX.17
Neue Reihe Nr 208

Lieber Herr Wittgenstein! Besten Dank für Ihre Karte vom 5.d.! Es freut mich immer sehr, ein Lebenszeichen von Ihnen zu erhalten und besonders eins, das von guter Stimmung zeugt. Ob mein nach Wien gerichteter Brief wohl in Ihre Hände gelangt ist? Ich konnte leider Ihre[r] so freundlichen Einladung nicht folgen. Ich wünsche Ihnen guten Erfolg Ihrer Arbeit und hoffe, nach Friedensschluss einmal mit Ihnen darüber sprechen zu können.

Herzlich grüsst Sie

Ihr ergebener G. Frege

Army Postcard

Prof. Frege from Jena
Brunshaupten in Mecklenb.
New Row Number 208

To Wittgenstein Fhrch
F. H. R. 5/4
Field Post 286
Austria-Hungary

Brunshaupten, 16.IX.17
New Row Number 208

Dear Mr. Wittgenstein, Many thanks for your card of the 5th! I am always very pleased to receive a sign of life from you and especially one testifying to your high spirits. Did my letter addressed to Vienna successfully reach you? I do regret that I could not accept your very kind invitation. I wish you much success in your work and hope to be able to speak with you about it after peace is concluded.

Kind regards to you,

Devotedly yours, G. Frege

Professor Dr. G. Frege
Neuburg bei Wismar

An Herrn Ludwig Wittgenstein
K. und K. Luftfahrtruppen
Flieger Kompagnie Nr 30
Feldpost Nr 470

26.II.18.

Beste Grüsse und Dank für Ihre Karte vom 8. dieses. Sie haben also, wie es scheint, schon wieder eine Verwandlung durchgemacht. Nun diese Tätigkeit mag sich ja mit Ihren früheren Studien gut in Einklang bringen lassen; denn, wenn ich mich recht erinnere, haben Sie sich mit Flugzeugen schon früher theoretisch beschäftigt. Und doch sehnen Sie sich nach der Beschäftigung mit viel tiefer liegenden Aufgaben, wie ich mir denken kann. Möge Ihnen ein baldiger Friedensschluss diesen Wunsch erfüllen und es Ihnen dann vergönnt sein, nicht nur für ein Reich, sondern für die Menschheit Grosses zu leisten. Dass wir uns dann einmal gesund wiedersehn, wünscht von Herzen

Ihr G. Frege

Army Postcard

Professor Dr. G. Frege,
Neuburg near Wismar

To Mr. Ludwig Wittgenstein
K. u. K. Aviation Squad
Pilot Company Number 30
Field Post Number 470

26.II.18.

Warmest regards and thanks for your card of the 8th of this month. It appears you have already undergone yet another metamorphosis. Now this current occupation certainly harmonizes well with your earlier studies; for, if I remember correctly at an earlier time you worked on the theory of airplanes. And yet you long to concern yourself with far deeper tasks, as I can well imagine. May a speedy conclusion of peace fulfill this wish and it be granted to you to achieve great things, not only for the sake of an empire, but for humanity.

Wishing, from the heart, that we may then someday see each other again in good health.

Yours, G. Frege

Neuburg bei Wismar, H. 21,
den 9. April 1918

Lieber Herr Wittgenstein!

Sie können sich kaum denken, mit welchem Erstaunen ich Ihren freundlichen Brief vom 25.III.18 gelesen habe. Sie schreiben von einer grossen Dankesschuld, die Sie mir gegenüber drückt. Ich weiss nichts von einer solchen. Jeder von uns, meine ich, hat vom Andern empfangen im geistigen Verkehr. Wenn ich mehr, als ich ahne, Sie in Ihren Bestrebungen gefördert habe, so freut mich das sehr; weiss ich doch, dass diese Bestrebungen in ihrem hohen Fluge die Welt der niedern Selbstsucht tief unter sich lassen. Was Sie in unsere[m] Verkehr gewonnen haben, das wird, hoffe ich, die Menschheit auf dem Wege, der ihr gewiesen ist, ein Stückchen vorwärts bringen. Wenn dabei die Worte, die ich mit Ihnen gewechselt habe, in ihren Wirkungen weiter leben werden, so ist das für mich ein tröstlicher Ausblick. Möge es Ihnen, lieber Freund, vergönnt sein, noch etwas von diesen Wirkungen zu erleben. Was bleibt mir übrig, als das, was Sie mir in edelster Regung zugedacht haben, mit herzlichem Danke anzunehmen, wie ich meine, dass es von Ihnen gemeint ist.

Mit freundschaftlichem Grusse

Ihr G. Frege.

Eine Benachrichtigung aus Jena ist mir noch nicht zugegangen; dagegen hat mir die Niederösterreichische Escompte-Gesellschaft die Akkreditierung bei der Bank für Thüringen angezeigt.

Letter

Neuburg near Wismar, H. 21,
9. April 1918

Dear Mr. Wittgenstein,

You can hardly imagine with what astonishment I read your friendly letter of 25.III.18. You write of the burden of a great debt of gratitude to me. I know of no such thing. Each of us, I believe, has drawn from the other in intellectual exchange. If I have helped you in your endeavors more than I suspect, that makes me very happy, as I know that the high flights of these endeavors leave the world of low self-interest far beneath them. I hope that whatever you have gained in our exchange will bring humanity a bit further forward along the road that has been pointed out to it. If as a result the words which I have exchanged with you shall live on in their effects, that is a consoling prospect for me. Dear friend, may it be granted to you to live to see some of these effects. What else remains but for me to accept what you, with most noble feeling, wanted me to have, with heartfelt thanks, as I think you intended.[6]

With friendly regards,

Yours, G. Frege

I have not yet received a notification from Jena; however, the Lower Austrian Escompte-Society has noted the credit to the Bank of Thuringia.

---

[6] Lothar Kreiser, Frege's biographer, reports that "Through a gift from Ludwig Wittgenstein at the beginning of 1918 and the sale of his Jena house in the same year it became possible for Frege to move back to his home region of Mecklenburg. Preparations for this began already in 1918. On 15.10.1918 Frege wrote for the first time under the name of his new place of residence, Bad Kleinen, to Wittgenstein" (*Gottlob Frege Leben-Werk-Zeit* (op. cit.) p. 504; see the letter to Wittgenstein of this date translated below). Kreiser adds that as a result of war loans and inflation the amount of Wittgenstein's gift to Frege was reduced, so that had Frege not owned the piece of land his house stood on in Bad Kleinen, with only his annual pension in reserve he "would have been standing at the threshold of poverty" (*Gottlob Frege Leben-Werk-Zeit*, p. 566).

An Wittgenstein Lt
K 1/ G 11
Fp. 386
Kanonenbatterie Nr 1 des K. u. K.
Gebirgsartillerieregiments Nr 11
Österreich-Ungarn

Neuburg b. Wismar, d. 1.VI.18.

Besten Dank für Ihre Karte vom 10.V. Es freut mich, dass Sie zu einem gewissen Abschlusse gekommen sind. Möge es Ihnen vergönnt sein, alles, was Sie herausgebracht haben, bald zu Papier zu bringen, damit es nicht verloren geht. Vielleicht werde auch ich dadurch gefördert in schwierigem Gelände, in dem ich mich abmühe. Natürlich bin ich immer bereit zu lernen und mich auf den richtigen Weg zurückführen zu lassen, wenn ich mich verirrt habe. Etwas Gewinn verspreche ich mir immer vom Kennenlernen der Wege, die Sie gegangen sind, auch dann, wenn ich im Wesentlichen Ihnen nicht zu folgen vermöchte. Glück auf zu weiterem rüstigen Schaffen!

Mit herzlichem Grusse

Ihr G. Frege

Army Postcard

To Lt. Wittgenstein
K 1/ G 11[7]
Fp. 386
Canon Battery Number 1 of the K. u. K.
Mountain Artillery Regiment Number 11
Austria-Hungary

Neuburg nr. Wismar, 1.VI.18.

Many thanks for your card of 10.V. I am pleased that you have arrived at a certain closure. May you soon be able to write down everything you have come up with so that it shall not be lost. This may also help me advance in the difficult area in which I am struggling. Naturally I am always ready to learn and be brought back onto the right path if I have gone astray. I always stand to gain from learning the paths which you have taken, even if I cannot follow you in essentials. Good luck with further strong work!

With kind regards,

Yours, G. Frege

---

[7] Frege's writing is difficult to read, but this line may say "K 1/ G 11". He left the return address blank on this card.

Herrn Ludwig Wittgenstein
Wien XVII
Neuwaldeggerstr. 38

Neuburg (Mecklenburg) d. 12.IX.18

L. H. W.! Eine Feldpostkarte, die ich schon vor längerer Zeit an Sie abgeschickt hatte, kam zurück mit der Bemerkung "im Spital". Nun wusste ich nicht, in welchem Spital, warum sie Ihnen nicht dahin nachgeschickt war und wie ich wieder mit Ihnen in Verbindung kommen könnte. Sie können sich denken, dass ich ernstlich um Sie besorgt war. Desto grösser ist nun meine Freude über Ihren lieben Brief, der heute in meine Hände gelangt ist. Ich glaube, aus ihm entnehmen zu können, dass Ihr Zustand wenigstens nicht besorgniserregend ist. Und besonders freue ich mich über das, was Sie über Ihre Arbeit schreiben. Es ist ja wohl begreiflich, dass einem, der sich selbst steile Steige zu bahnen sucht, wo noch kein Mensch vor ihm gewesen ist, manchmal die Frage nahe tritt, ob nicht vielleicht alles vergeblich sei, ob irgend jemand jemals Lust haben werde, diesen Steigen nachzugehen. Ich kenne das auch; aber ich habe doch jetzt die Zuversicht, dass nicht alles vergeblich gewesen ist. Durch Übersendung eines Exemplars würden Sie mich sehr erfreuen. Ich denke, dass von mir demnächst eine Kleinigkeit erscheinen wird, die ich Ihnen als Gegengabe zugehen lassen kann. Es wird vielleicht wenig Neues darin sein; aber doch vielleicht in neuer Weise gesagt und dadurch manchem verständlicher. Hoffentlich wird es mir vergönnt sein, im mündlichen Verkehr Ihre Ansichten vollkommener kennen zu lernen, als es durch das gedruckte Wort möglich ist. Unsere Spaziergänge in Jena und Brunshaupten sind mir noch immer in schöner Erinnerung. In alter Freundschaft

herzlich grüssend Ihr G. Frege

Postcard

Mr. Ludwig Wittgenstein
Vienna XVII
Neuwaldeggerstr. 38

Neuburg (Mecklenburg) 12.IX.18

D.M.W.! An army postcard which I had sent you long ago came back with the remark "In the hospital". Now I did not know which hospital, why it was not sent to you there, and how I could get back in touch with you. You can imagine that I was very concerned about you. How much greater my pleasure, then, in receiving your kind letter which came to me today. I gather from it that your condition is at least no cause for worry. And I am especially pleased about what you write of your work. It is indeed understandable that someone who tries on his own to clear steep paths where no other person has been before him sometimes broaches the question whether everything might not be futile, whether anyone will ever be inclined to follow this path. I know this too; but I now have confidence that not everything has been futile. By sending a copy you would make me very glad. I think that shortly a little something will appear from me, which I intend to send you as a return gift. There is perhaps little new in it; but perhaps said in a new way and therefore more intelligible to some. I hope I may have the privilege of learning about your views more thoroughly in conversation than is possible via the printed word. Our walks in Jena and Brunshaupten are still a beautiful memory for me. In abiding friendship

Kind regards, Yours, G. Frege

Dr. G. Frege, Professor
Bad Kleinen Mecklenb. N 52

An L. Wittgenstein Lt.
G.A.R.11 Bt. 1
Feldpost 280[8] Österreich-Ungarn

Bad Kleinen Mecklenb. N. 52, den 15.X.18

Durch Ihre Karte haben Sie mich sehr erfreut. Ich beglückwünsche Sie zu dem Abschluss Ihrer Arbeit und bewundere Sie, dass Sie es in dieser Zeit und unter solchen Umständen fertig gebracht haben. Möchte es Ihnen vergönnt sein, die Arbeit gedruckt zu sehen, und mir, sie zu lesen! Ich hoffe Ihnen nächstens etwas von mir senden zu können. Wahrscheinlich werden Sie nicht gerade damit einverstanden sein; aber desto anregender werden wir darüber sprechen können, wenn es uns vergönnt sein sollte, uns in freundlicheren und friedlicheren Zeiten gesund wiederzusehen. Schon habe ich eine zweite kleine Abhandlung über die Verneinung im Rohen fertig, die ich dann, sobald es geht, zu veröffentlichen gedenke. Sie ist als Fortsetzung der ersten gedacht. Ich danke Ihnen für Ihr treues Gedenken und werde auch Ihrer immer in Freundschaft gedenken.

Auch ich hoffe auf ein Wiedersehen.

In herzlicher Freundschaft

Ihr G. Frege

[8] The address was double underlined and crossed out in the original.

Army Postcard

Dr. G. Frege, Professor
Bad Kleinen Mecklenb. N 52

To L. Wittgenstein, Lt.
G.A.R.11 Bt. 1
Field Post 280 Austria-Hungary[9]

Bad Kleinen Mecklenb. Nr. 52, 15.X.18

Your card pleased me very much. I congratulate you on the conclusion of your work and admire you for having managed it during these times and under such conditions. May you see the work in print, and may I read it! I hope to be able to send you something soon.[10] You will probably not agree with it entirely; but then there will be even more of an edge to our talks about it when we see each other healthy again, in friendlier and more peaceful times. I have already finished a draft of a second small treatise on negation that I am thinking of publishing as soon as can be arranged.[11] I conceive of it as the continuation of the first. I am grateful for your faithful remembrance of me; I will always think of you too in heartfelt friendship.

I too hope to see you again.

In affectionate friendship

Yours, G. Frege.

---

[9] In Frege's hand the address line reads: "G.A.R. 11 Bt. 1/Field Post 379", but someone crossed out the last number and wrote instead: "8/G.A.R. 11" and then changed the Field Post number to 280. Compare Wittgenstein's letter to Engelmann of 9 April 1918, in Paul Engelmann, *Letters from Ludwig Wittgenstein, With a Memoir* (Oxford: Basil Blackwells, 1967/New York: Horizon Press, 1968), also on the Intelex CD-ROM of Wittgenstein, *Briefwechsel*.

[10] Frege is referring here to his essay "Der Gedanke. Eine logische Untersuchung", in *Beiträge zur Philosophie des deutschen Idealismus*, vol. I, pp. 58–77, to which he also refers in his letter to Wittgenstein of 16 September 1919. An English translation of this essay appears under the title "Thoughts" in Frege, *Collected Papers on Mathematics, Logic, and Philosophy*, ed. B. McGuinness, trans. M. Black et al. (Oxford: Basil Blackwell, 1984), pp. 351–372.

[11] "Die Verneinung", published in *Beiträge zur Philosophie des deutschen Idealismus* I (1918–1919), pp. 143–157; translated into English as "Negation", in Frege, *Collected Papers on Mathematics, Logic, and Philosophy* (op. cit.), pp. 373–389.

Bad Kleinen in Mecklenburg, den 28.VI.19

Lieber Freund!

Sie warten gewiss schon längst auf eine Antwort von mir und erwünschen eine Äusserung von mir über Ihre Abhandlung, die Sie mir haben zukommen lassen. Ich fühle mich deshalb sehr in Ihrer Schuld und hoffe auf Ihre Nachsicht. Ich bin in der letzten Zeit sehr mit langwierigen geschäftlichen Angelegenheiten belastet gewesen, die mir viel Zeit weggenommen haben, weil ich in der Erledigung solcher Sachen aus Mangel an Übung ungewandt bin. Dadurch bin ich verhindert worden, mich mit Ihrer Abhandlung eingehender zu beschäftigen und kann daher leider Ihnen kein begründetes Urteil darüber abgeben. Ich finde sie schwer verständlich. Sie setzen Ihre Sätze nebeneinander meistens ohne sie zu begründen oder wenigstens ohne sie ausführlich genug zu begründen. So weiss ich oft nicht, ob ich zustimmen soll, weil mir der Sinn nicht deutlich genug ist. Aus einer eingehenderen Begründung würde auch der Sinn klarer hervorgehen. Der Sprachgebrauch des Lebens ist im Allgemeinen zu schwankend, um ohne Weiteres für schwierige logische und erkenntnistheoretische Zwecke brauchbar zu sein. Es sind, wie mir scheint, Erläuterungen nötig, um den Sinn schärfer auszuprägen. Sie gebrauchen gleich am Anfange ziemlich viele Wörter, auf deren Sinn offenbar viel ankommt.

Gleich zu Anfang treffe ich die Ausdrücke "der Fall sein" und "Tatsache" und ich vermute, dass *der Fall sein* und *eine Tatsache sein* dasselbe ist. Die Welt ist alles, was der Fall ist und die Welt ist die Gesamtheit der Tatsachen. Ist nicht jede Tatsache der Fall und ist nicht, was der Fall ist, eine Tatsache? Ist [es] nicht dasselbe, wenn ich sage, A sei eine Tatsache, wie wenn ich sage, A sei der Fall? Wozu dieser doppelte Ausdruck? Freilich ist jedes gleichseitige Dreieck ein gleichwinkliges Dreieck und jedes gleichwinklige Dreieck ein gleichseitiges Dreieck und doch ist der Sinn des ersten Ausdrucks nicht zusammenfallend mit dem des zweiten. Es ist ein Lehrsatz[,] dass jedes gleichseitige Dreieck ein gleichwinkliges ist. Aber hier sind die Ausdrücke "gleichseitiges Dreieck" und "gleich[winkliges] Dreieck" zusammengesetzt und aus der verschiedenen Zusammensetzung ergibt sich ein verschiedener Sinn. Aber in unserm Falle haben wir das nicht. Kann man sagen, aus der Zusammensetzung des Ausdrucks "der Fall sein" ergebe sich der Sinn? Ist es ein Lehrsatz, dass, was der Fall ist, eine Tatsache ist?

Letter

Bad Kleinen in Mecklenburg, 28.VI.19

Dear friend,

You have certainly long awaited an answer from me and must want me to comment on your treatise that you sent to me. On this account I feel myself very indebted to you and hope for your indulgence. I have recently been burdened with many tedious business affairs that have taken much time away from me because I am not accustomed to handling such matters. I have thus been prevented from occupying myself more thoroughly with your treatise and can therefore unfortunately give you no well-grounded judgment. I find it difficult to understand. For the most part you put your sentences down one beside the other without justification, or at least without sufficiently detailed justification. I thus often do not know whether I ought to agree, for their sense is not sufficiently clear to me. Surely the sense would become clearer with more detailed justification. In general colloquial language is too faltering to be suited, just as it is, for difficult logical and epistemological tasks. It seems to me that elucidations are necessary to make the sense more precise. Right from the beginning you use many words upon whose senses evidently much depends.

Right at the beginning I encounter the expressions "to be the case" and "fact" and I conjecture that *to be the case* and *to be a fact* are the same. The world is everything that is the case and the world is the totality of facts. Is not every fact the case, and is not what is the case a fact? Is it not the same when I say, Let A be a fact, as when I say, Let A be the case? What is the point of this double expression? Of course, every equilateral triangle is an equiangular triangle and every equiangular triangle is an equilateral triangle, but nevertheless the sense of the first expression does not coincide with that of the second. It is a theorem that every equilateral triangle is also an equiangular one. But here the expressions "equilateral triangle" and "equiangular triangle"[12] are composite, and from their different compositions result different senses. However, in our case we do not have this. Can one say that the sense of the expression "to be the case" results from its composition? Is it a theorem that what is the case, is a fact?

[12] Frege actually wrote "equilateral triangle" in this second-quoted phrase, the same as what he had written in the first-quoted phrase, but presumably this was a slip of the pen. We assume that he intended to write "equiangular triangle" here, for only then would he be able to say, as he does, that the first- and second-quoted expressions have "different compositions" and hence "different senses".

Ich meine nicht; aber auch als Axiom möchte ich es nicht gelten lassen; denn irgendeine Erkenntnis scheint mir darin nicht zu liegen. Nun kommt aber noch ein dritter Ausdruck: "Was der Fall ist, die Tatsache, ist das Bestehen von Sachverhalten." Ich verstehe das so, dass jede Tatsache das Bestehen eines Sachverhaltes ist, so dass eine andre Tatsache das Bestehen eines andern Sachverhaltes ist.

Könnte man nun nicht die Worte "das Bestehen" streichen und sagen: "Jede Tatsache ist ein Sachverhalt, jede andre Tatsache ist ein anderer Sachverhalt." Könnte man vielleicht auch sagen "Jeder Sachverhalt ist das Bestehen einer Tatsache"? Sie sehen: ich verfange mich gleich anfangs in Zweifel über das, was Sie sagen wollen, und komme so nicht recht vorwärts. Ich fühle mich jetzt oft müde, und das erschwert mir das Verständnis gleichfalls. Sie werden, hoffe ich, mir diese Bemerkungen nicht verübeln, sondern sie als Anregung betrachten, die Ausdrucksweise in Ihrer Abhandlung verständlicher zu machen. Wo so viel auf genaue Erfassung des Sinnes ankommt, darf man dem Leser nicht zu viel zumuten. An sich scheint mir der Gebrauch verschiedener Ausdrücke in demselben Sinne ein Übel zu sein, wo man es besonderer Vorteile wegen doch tut, sollte man den Leser darüber nicht im Zweifel lassen. Wo aber der Leser wider die Absicht des Schriftstellers dazu kommen könnte mit verschiedenen Ausdrücken denselben Sinn zu verbinden, sollte der Schriftsteller auf die Verschiedenheit hinweisen und möglichst deutlich zu machen suchen, worin sie besteht. Gibt es auch Sachverhalte, die nicht bestehen? Ist jede Verbindung von Gegenständen ein Sachverhalt? Kommt es nicht auch darauf an, wodurch diese Verbindung hergestellt wird? Was ist das Verbindende? Kann dieses vielleicht die Gravitation sein wie beim Planetensystem? Ist dieses ein Sachverhalt? Sie schreiben: "Es ist für das Ding wesentlich, der Bestandteil eines Sachverhaltes sein zu können." Kann nun ein Ding auch Bestandteil einer Tatsache sein? Der Teil des Teils ist Teil des Ganzen. Wenn ein Ding Bestandteil einer Tatsache ist und jede Tatsache Teil der Welt ist, so ist auch das Ding Teil der Welt. Zum besseren Verständnisse wünsche ich Beispiele, schon, um zu sehen, was sprachlich der Tatsache, dem Sachverhalte, der Sachlage entspricht, wie sprachlich eine Tatsache, ein bestehender und etwa ein nicht bestehender Sachverhalt bezeichnet wird und wie das Bestehen eines Sachverhaltes und also die dem Sachverhalte entsprechende Tatsache bezeichnet wird, ob sich dabei ein wesentlicher Unterschied zwischen einem Sachverhalte und der Tatsache ergibt. Ein Beispiel möchte ich haben dafür, dass der Vesuv Bestandteil eines Sachverhaltes ist. Dann müssen, wie es scheint, auch Bestandteile des Vesuvs Bestandteile dieser Tatsache sein; die Tatsache wird also auch aus erstarrten Laven bestehen. Das will mir nicht recht scheinen.

I think not; however I would also not consider it an axiom; for it seems to me that no knowledge lies within it. However, still a third expression now appears: "What is the case, the fact, is the existence of atomic facts."[13] I understand this to say that every fact is the existence of an atomic fact, so that a distinct fact is the existence of a distinct atomic fact.

Could one not now strike out the words "the existence of" and say, "Every fact is an atomic fact, every distinct fact is a distinct atomic fact." Could one perhaps also say, "Every atomic fact is the existence of a fact"? You see: I am entangled from the very beginning in doubts about what you mean to say, and thus I make no progress. I often feel tired now, and this also makes it difficult for me to comprehend. I hope you will not take these remarks amiss, but, rather, consider them as an encouragement to make the manner of expression in your treatise easier to understand. Where so much depends on a precise grasp of the sense, one must not expect too much of the reader. It seems to me that the use of different expressions with the same sense is in itself an evil. Where one still does this in order to gain something special, one should not leave the reader in doubt about it.Where the reader could, however, in contradiction to the intention of the writer, combine two different expressions with the same sense, the writer should point out the difference and seek to make as clear as possible wherein it lies. Are there also atomic facts that do not exist? Is every combination of objects an atomic fact? Isn't it of importance by what means these combinations are produced? What is the thing that binds? Can this perhaps be gravitation, as with the system of planets? Is this an atomic fact? You write: "It is essential to a thing that it can be a constituent part of an atomic fact." Can now a thing also be a constituent of a fact? The part of a part is part of the whole. If a thing is a constituent of a fact and every fact part of the world, then also the thing is part of the world. For the sake of better understanding I would like to have examples, in order to see what corresponds linguistically to the fact, to the atomic fact, to the state of affairs; how a fact, an existing and a non-existing atomic fact are linguistically signified and how the existing of an atomic fact and, hence, the fact corresponding to the atomic fact are signified; whether thereby an essential distinction results between an atomic fact and the fact.[14] I want to have an example for the claim that Vesuvius is a constituent of an atomic fact. Then it appears that constituents of Vesuvius must also be constituents of this fact; the fact will therefore also consist of hardened lava. That does not seem right to me.

---

[13] Here we have translated "*Sachverhalt*" as "atomic fact" and "*Sachlage*" as "state of affairs", following the earlier translation of Wittgenstein's *Tractatus* by C.K. Ogden, a translation that Wittgenstein himself worked through and accepted.

[14] Here we eliminate a "." written in Frege's hand, reading for it "*und*", according to a suggestion of Christian Thiel.

Doch ich wollte Ihnen ja mit diesen Zeilen einen Freundschaftsdienst erweisen und nun fürchte ich, Sie mit zudringlichen Fragen belästigt zu haben. Verzeihen Sie dies und bewahren Sie die Freundschaft

Ihrem oft an Sie denkenden

G. Frege

I do however want these lines to prove a service of friendship to you and now I fear that I have harassed you with intrusive questions. Forgive this and do keep your friendship toward someone who is

thinking of you often, Your

G. Frege

Bad Kleinen, den 16. Sept. 1919.

Lieber Herr Wittgenstein!

Noch habe ich Ihren letzten Brief aus Cassino nicht beantwortet, da erhalte ich schon wieder ein Schreiben von Ihnen. Vielen Dank für beide. Ich beglückwünsche Sie zu der glücklichen Rückkehr aus der Gefangenschaft. Mögen Sie bald die Folgen alles dessen überwinden, was Sie haben durchmachen müssen. Dass Sie einen Beruf ergreifen wollen, scheint mir erfreulich, und ich wünsche Ihnen herzlich, dass die Hoffnungen, die Sie damit verbinden, sich erfüllen werden. Ich halte die Aussicht, dass wir uns auf philosophischem Gebiete noch verständigen werden, nicht für so gering, wie Sie es zu tun scheinen. Ich verbinde damit die Hoffnung, dass Sie dereinst für das, was ich im Gebiete der Logik erkannt zu haben glaube, eintreten werden. Zuvor müssten Sie freilich dafür gewonnen werden. Deswegen ist mir der Meinungsaustausch mit Ihnen erwünscht. Und ich habe in langen Gesprächen mit Ihnen einen Mann kennen gelernt, der gleich mir nach der Wahrheit gesucht hat, z. Tl auf andern Wegen. Aber gerade dies lässt mich hoffen, bei Ihnen etwas zu finden, was das von mir Gefundene ergänzen, vielleicht auch berichtigen kann. So erwarte ich, indem ich versuche, Sie zu lehren, mit meinen Augen zu sehen, selbst zu lernen, mit Ihren Augen zu sehen. Die Hoffnung auf eine Verständigung mit Ihnen gebe ich so leicht nicht auf.

Erlauben Sie, dass ich auf den Inhalt Ihres letzten Briefes noch nicht eingehe. Schon der vorige aus Cassino hat soviel in mir in Bewegung gesetzt, dass, wenn ich allen Anregungen Folge geben wollte, eher ein Buch als ein Brief entstände.

Was Sie mir über den Zweck Ihres Buches schreiben, ist mir befremdlich. Danach kann er nur erreicht werden, wenn Andere die darin ausgedrückten Gedanken schon gedacht haben. Die Freude beim Lesen Ihres Buches kann also nicht mehr durch den schon bekannten Inhalt, sondern nur durch die Form erregt werden, in der sich etwa die Eigenart des Verfassers ausprägt. Dadurch wird das Buch eher eine künstlerische als eine wissenschaftliche Leistung; das, was darin gesagt wird, tritt zurück hinter das, wie es gesagt wird. Ich ging bei meinen Bemerkungen von der Annahme aus, Sie wollten einen neuen Inhalt mitteilen. Und dann wäre allerdings grösste Deutlichkeit grösste Schönheit.

Ob ich zu denen gehöre, die Ihr Buch verstehen werden? Ohne Ihre Beihülfe schwerlich. Auf das, was Sie mir über Sachverhalt, Tatsache, Sachlage schreiben, wäre ich von selbst kaum verfallen, wiewohl ich an einer Stelle meines Aufsatzes[15] Ihrer Meinung wohl nahe komme.

---

[15] Der Gedanke, eine logische Untersuchung, in den *Beiträgen zur Philosophie des Deutschen Idealismus*, I. Bd. S. 58.

Letter

Bad Kleinen, 16. Sept. 1919.

Dear Mr. Wittgenstein,

I still have not answered your last letter from Cassino, and now I have already received another letter from you. Many thanks for both. I congratulate you on your safe return from imprisonment. May you soon overcome the consequences of everything you had to go through. I am pleased that you want to take up a profession, and I sincerely wish that the hopes you place in it will be fulfilled. In my view it is not as unlikely as it seems to you that we might still manage to come to agreement in the philosophical domain. I combine this with the hope that you will some day advance what I believe I have discovered in the domain of logic. First, however, you must be well won over to it. That is why I would like to exchange opinions with you. In long conversations with you I have come to know a man who, like me, has sought the truth, partly on different paths. But this is precisely what gives me the hope of finding something in you that can supplement what I have found, perhaps even correcting it.[16] So I expect that while trying to teach you to see with my eyes, I shall be learning to see with your eyes. I shall not so easily surrender the hope of reaching agreement with you.

Allow me not yet to enter into discussion of the content of your most recent letter. Already the earlier one from Cassino set so much in motion in me that if I followed up on every stimulating point I would have to write a book rather than a letter.

What you write me about the purpose of your book strikes me as strange. According to you, that purpose can only be achieved if others have already thought the thoughts expressed in it. The pleasure of reading your book can therefore no longer arise through the already known content, but, rather, only through the form, in which is revealed something of the individuality of the author. Thereby the book becomes an artistic rather than a scientific achievement; that which is said therein steps back behind how it is said. I had supposed in my remarks that you wanted to communicate a new content. And then the greatest distinctness would indeed be the greatest beauty.

Am I one of those who will understand your book? Without your assistance, hardly. What you write me about atomic facts, facts, and states of affairs would never have occurred to me, although possibly I come close to your opinion at one place in my essay.[17]

---

[16] Frege's handwriting is difficult to read here: he may have written "*noch*" or "*auch*" in this sentence. If the latter, then the translation would read "perhaps also setting it right".

[17] Frege adds a footnote here: "Der Gedanke, eine Logische Untersuchung, in den *Beiträgen zur Philosophie des Deutschen Idealismus*, I Bd. S. 58."

Um so mehr freut es mich in Ihrem Briefe einen Satz zu finden, in dem Ihre Sprechweise ganz mit meiner eigenen übereinzustimmen scheint. Es ist der Satz: "Der Sinn jener beiden Sätze ist ein und derselbe, aber nicht die Vorstellungen, die ich mit ihnen verband, als ich sie schrieb." Hier stimme ich Ihnen ganz bei, dass Sie den Satz von seinem Sinne unterscheiden, die Möglichkeit offen lassend, dass zwei Sätze denselben Sinn haben und sich dann noch durch Vorstellungen unterscheiden, die mit ihnen verbunden werden. In dem unten genannten Aufsatze habe ich auf S. 63 davon gehandelt. Sie unterstreichen das Wort "ich". Auch darin sehe ich ein Zeichen der Übereinstimmung. Der eigentliche Sinn des Satzes ist für Alle derselbe; die Vorstellungen aber, die jemand mit dem Satze verbindet, gehören ihm allein an; er ist ihr Träger. Niemand kann die Vorstellungen eines Andern haben.

Sie schreiben nun: "Was einem Elementarsatze entspricht, wenn er wahr ist, ist das Bestehen eines Sachverhaltes." Hiermit erklären Sie nicht den Ausdruck "Sachverhalt", sonder[n] den ganzen Ausdruck, "das Bestehen eines Sachverhaltes". In einer Definition muss der erklärte Ausdruck immer als untrennbar Ganzes angesehen werden. Die Teile, die man grammatisch in ihm unterscheiden kann, sind nicht als solche aufzufassen, die einen eigenen Sinn haben. Sie gebrauchen das Wort "Bestehen" auch in anderen Zusammenhängen. Danach scheint sich Ihnen der Ausdruck "das Bestehen eines Sachverhaltes" in zwei Teile zerlegt zu haben, und Ihr Satz "Was einem Elementarsatze entspricht, wenn er wahr ist, ist das Bestehen eines Sachverhaltes" scheint nicht eine Erklärung des Ausdrucks "das Bestehen eines Sachverhaltes" zu sein. Ich versuche Ihre Erklärung so aufzufassen: "Ein Elementarsatz kann ohne Änderung des Sinnes umgeformt werden in einen Satz von der Form ≪A besteht≫.["] Hierbei ist der Sinn des Wortes "besteht" als bekannt angenommen. Wenn nun jener Elementarsatz wahr ist, so ist A ein Sachverhalt. Danach kann man auch sagen: "Wenn der Satz ≪A besteht≫ ein wahrer Elementarsatz ist, so ist A ein Sachverhalt"; denn dieser Satz braucht nicht erst umgeformt zu werden, da er die geforderte Form schon hat.

Doch nun muss ich wohl erst abwarten, was Sie dazu sagen.

Seien Sie bis dahin der Freundschaft versichert Ihres ergebenen

G. Frege.

I am all the more pleased to find in your letter a sentence in which your way of speaking seems to be entirely in agreement with mine. It is the sentence: "The sense of each of the two sentences is one and the same, but not the ideas which I combined with them when I wrote them." Here I fully agree with your distinguishing the sentence from its sense, leaving open the possibility that two sentences have the same sense and yet may still differ in the ideas which are combined with them. In the aforementioned essay I have treated the matter on p. 63. You underline the word "I". In this too I see a sign of agreement. The real sense of the sentence is the same for everyone; however, the ideas which someone combines with the sentence belong to him alone; he is their bearer. No one can have the ideas of someone else.

You now write: "What corresponds to an elementary proposition, if it is true, is the existence of an atomic fact." With this you explain, not the expression "atomic fact", but rather the whole expression, "the existence of an atomic fact". In a definition the expression explained must always be viewed as an inseparable whole. The parts which one can distinguish in it grammatically are not to be conceived as having their own senses. You use the word "existing" in other contexts as well. Thus in your hands the expression "the existence of an atomic fact" appears to have been divided into two parts, and your sentence "What corresponds to an elementary proposition, if it is true, is the existence of an atomic fact" appears not to be an explanation of the expression "the existence of an atomic fact". I try to grasp your explanation this way: "An elementary proposition can be transformed into a sentence of the form 'A exists' without change of sense." Here the sense of the word "exists" is supposed known. If now that elementary proposition is true, then A is an atomic fact. Thus one can also say: "If the proposition 'A exists' is a true elementary proposition, then A is an atomic fact"; for this sentence does not need to be first transformed for it already has the required form.

But now I must wait to hear what you say about this.

Be assured until then of the friendship of your devoted

G. Frege

Bad Kleinen i. Mecklenb., den 30.IX.19

Lieber Herr Wittgenstein!

Ihre Bitte, Ihnen zum Drucke Ihrer Abhandlung in den Beiträgen z. Ph. d. D. I. behilflich zu sein, habe ich mir durch den Kopf gehen lassen. Ich kenne von den Herren persönlich nur Prof. Bauch in Jena. Über die Aufnahme eines Beitrages entscheidet, wie mir scheint, meist Herr Hoffmann in Erfurt allein. Für diesen würde aber, wie ich glaube, eine Empfehlung von Prof. Bauch von entscheidendem Einflusse sein. Soll ich mich an diesen wenden? Ich könnte ihm schreiben, dass ich sie als durchaus ernst zu nehmenden Denker kennen gelernt habe. Über die Abhandlung selbst kann ich kein Urteil abgeben, nicht, weil ich mit dem Inhalte nicht einverstanden bin, sondern, weil mir der Inhalt zu wenig klar ist. Vielleicht würden wir, nachdem wir uns erst einmal über den Wortgebrauch verständigt hätten, finden, dass wir garnicht sehr voneinander abweichen. Ich könnte bei Prof. Bauch anfragen, ob er das Mscrpt zu sehen wünsche. Ich glaube aber kaum, dass dies einen Erfolg haben würde. Wenn ich mich nicht verrechnet habe, würde Ihr Mscrpt etwa 50 Seiten der Beiträge füllen, also vielleicht in einem Hefte der Beiträge grade Platz finden. Es scheint mir aussichtslos, dass der Herausgeber ein ganzes Heft einem einzigen, noch dazu unbekannten Schriftsteller einräume. Wenn an eine Veröffentlichung in einer Zeitschrift gedacht werden soll, dürfte eine Zerteilung der Abhandlung nötig sein. Sie schreiben in Ihrem Vorworte, dass Ihnen die Wahrheit der mitgeteilten Gedanken unantastbar und definitiv scheine. Könnte nun nicht einer dieser Gedanken, in dem die Lösung eines philosoph. Problems enthalten ist, zum Gegenstande einer Abhandlung genommen werden und so das Ganze in soviele Teile zerlegt werden, als philosoph. Probleme behandelt werden? Es ist auch gut, den Leser nicht durch die Länge der Abhandlung kopfscheu zu machen. Wenn die erste Abhandlung, die das Grundlegende enthalten müsste, Anklang fände, wäre es leichter auch die übrigen Abhandlungen in der Zeitschrift unterzubringen. Dabei könnte vielleicht noch ein Übelstand vermieden werden. Nachdem man Ihr Vorwort gelesen hat, weiss man nicht recht, was man mit Ihren ersten Sätzen anfangen soll. Man erwartet eine Frage, ein Problem gestellt zu sehen und nun liest man etwas, was den Eindruck von Behauptungen macht, die ohne Begründungen gegeben werden, deren sie doch dringend bedürftig erscheinen. Wie kommen Sie zu diesen Behauptungen? Mit welchem Probleme hängen sie zusammen? Ich möchte eine Frage an die Spitze gestellt sehen, ein Rätsel, dessen Lösung kennen zu lernen, erfreuen könnte. Man muss gleich anfangs Mut schöpfen, sich mit dem Folgenden zu befassen.

Letter

Bad Kleinen i. Mecklenb., 30.IX.19

Dear Mr. Wittgenstein,

I have been thinking over your request for help in publishing your treatise in the *Beiträge z. Ph. d. D. I.*[18] Of the gentlemen involved, Professor Bauch in Jena is the only one I know personally. I am under the impression that the decision about accepting a submitted work usually rests solely with Mr. Hoffmann in Erfurt. For him, however, a recommendation from Professor Bauch would have, I believe, decisive influence. Should I turn to him? I could write to him that I have come to know you as a thinker to be taken rather seriously. About the treatise itself I can render no judgment, not because I am not in agreement with the content, but because the content is not sufficiently clear to me. If we were only able to reach agreement about the use of words perhaps we would find that we do not differ with one another substantially. I could ask Professor Bauch whether he would like to see the manuscript. I am doubtful, however, that this would be successful. If I have not miscalculated, your manuscript would fill around 50 pages of the *Beiträge*, thus perhaps taking up an entire volume of the *Beiträge*. I see no prospect that the editor would give the space of an entire volume to a single writer, and an unknown one at that. If publication in a journal is to be considered, the treatise would have to be divided into parts. You write in your Preface that the truth of the thoughts communicated seems to you unassailable and definitive. Could not then one of these thoughts, in which the solution of a philosophical problem is contained, be taken as the object of a treatise, so that the whole would be divided into so many parts, just as philosophical problems are treated? It is also better not to intimidate the reader with the length of the treatise. If the first treatise, which would have to contain the fundamentals, were well received, it would be easier to place the remaining treatises in the journal. Moreover, in this way another hindrance might be avoided. Having read your Preface, one does not really know what one is supposed to do with your first propositions. One expects to see a question, a problem posed, and instead one reads what appear to be assertions, in urgent need of justification, but given with none. How do you arrive at these assertions? With which problems are they connected? I would like to see a question posed at the beginning, a riddle whose solution one would be pleased to know. One must be given courage from the very beginning to be able to deal with what follows.

[18] *Beiträge zur Philosophie des deutschen Idealismus.* Frege did contact both Bauch and Hoffmann, who are mentioned here; see G. Gabriel et al., eds., Gottlob Frege, *Wissenschaftlicher Briefwechsel*, vol. II of *Nachgelassene Schriften und Wissenschaftlicher Briefwechsel*, (Hamburg: Felix Meiner Verlag, 1976), pp. 8–9, 81, including the notes concerning letters from Bauch dated 31 October 1919 (III/4) and Hoffmann dated 23 January 1920 (XVI/3).

Doch sind das im Grunde Fragen, die Sie nur selbst beantworten können. Es fehlt mir eine eigentliche Einleitung, in der ein Ziel gesteckt wird.

Doch nehmen Sie mir diese Bemerkungen nicht übel; sie sind in guter Absicht gemacht.

Bewahren Sie vielmehr Ihre Freundschaft

Ihrem G. Frege.

But these are at bottom questions which you alone can answer. I miss a genuine introduction in which an aim is set.

But do not take offense at my remarks; they are made with good intentions.

On the contrary, do keep up your friendship toward

Your G. Frege

Bad Kleinen (Mecklenb.), den 3.IV.20

Lieber Herr Wittgenstein!

Besten Dank für Ihren Brief vom 19.III! Natürlich nehme ich Ihnen Ihre Offenheit nicht übel. Aber ich möchte gerne wissen, welche tieferen Gründe des Idealismus Sie meinen, die ich nicht erfasst hätte. Ich glaube verstanden zu haben, dass Sie selbst den erkenntnistheoretischen Idealismus nicht für wahr halten. Damit erkennen Sie, meine ich, an, dass es tiefere Gründe für diesen Idealismus überhaupt nicht gibt. Die Gründe dafür können dann nur Scheingründe sein, nicht logische. Man wird ja zuweilen von der Sprache irre geführt, weil die Sprache nicht immer den logischen Ansprüchen genügt. Bei der Bildung der Sprache ist ja neben den logischen Fähigkeiten des Menschen sehr viel Psychologisches wirksam gewesen. Logische Fehler stammen nicht aus der Logik, sondern kommen von den Verunreinigungen oder Störungen her, denen die logische Tätigkeit des Menschen ausgesetzt ist. Meine Absicht war es nicht, allen solchen Störungen psychologisch-sprachlicher Herkunft nachzuspüren. Gehen Sie, bitte, einmal meinen Aufsatz über den Gedanken durch bis zu dem ersten Satze, dem Sie nicht zustimmen, und schreiben Sie mir diesen Satz und die Gründe Ihrer Abweichung. So werde ich wohl am besten erkennen, was Sie im Auge haben. Vielleicht habe ich garnicht in dem Sinne, wie Sie es meinen, den Idealismus bekämpfen wollen. Ich habe den Ausdruck "Idealismus" überhaupt wohl nicht gebraucht. Nehmen Sie meine Sätze ganz, wie sie dastehen, ohne mir eine Absicht unterzuschieben, die mir vielleicht fremd gewesen ist.

Was nun Ihre eigene Schrift anbetrifft, so nehme ich gleich an dem ersten Satze Anstoss. Nicht, dass ich ihn für falsch hielte, sondern weil mir der Sinn unklar ist. "Die Welt ist alles, was der Fall ist." Das "ist" wird entweder als blosse Copula gebraucht, oder wie das Gleichheitszeichen in dem volleren Sinne von "ist dasselbe wie". Während das "ist" des Nebensatzes offenbar blosse Copula ist, kann ich das "ist" des Hauptsatzes nur in dem Sinne eines Gleichheitszeichens verstehen. Bis hier ist, glaube ich, kein Zweifel möglich. Aber ist die Gleichung als Definition zu verstehen? Das ist nicht so deutlich. Wollen Sie sagen: "Ich will unter ≪Welt≫ verstehen alles, was der Fall ist"? Dann ist "die Welt" der erklärte Ausdruck, "alles was der Fall ist" der erklärende. In diesem Falle wird nichts damit behauptet von der Welt oder von dem, was der Fall ist, sondern, wenn etwas behauptet werden soll, so ist es etwas über den Sprachgebrauch des Schriftstellers. Ob und wieweit dieser etwa mit dem Sprachgebrauch des Lebens übereinstimme, ist eine Sache für sich, auf die aber für den Philosophen wenig ankommt, nachdem er seinen Sprachgebrauch einmal festgestellt hat.

Aber in einer Gleichung kann auch ein Gedanke ausgedrückt werden, der unsere Erkenntnis wesentlich erweitert, wenn er als wahr anerkannt wird. Jede Wiedererkennung ist eine Erkenntnis dieser Art.

Letter

Bad Kleinen (Mecklenb.), 3.IV.20

Dear Mr. Wittgenstein,

Many thanks for your letter of March 19! Of course I do not take offense at your frankness. But I would like to know which deeper grounds of idealism you think I have not grasped. I believe that I understood that you yourself do not hold epistemological idealism to be true. Therefore you acknowledge, I think, that there are no deeper grounds for this idealism at all. The grounds for it can then only be apparent, not logical. Yes, one is sometimes led astray by language, because language is not always up to the demands of logic. Indeed, in the formation of language a great deal that is psychological was at work alongside the logical capacities of humankind. Logical mistakes do not stem from logic, but arise from the impurities or disturbances to which the logical activity of a human being is subjected. It was not my intention to trace all such disturbances of psychologico-linguistic origin. Would you please go through my essay on thought until the first sentence with which you disagree, and write me this sentence along with the reasons why you disagree. Probably that will be the best way for me to find out what you have in mind. Perhaps I did not want to fight idealism in the sense in which you mean it. I probably did not even use the expression "idealism" at all. Take my sentences just as they stand, without attributing to me any intention that might have been foreign to me.

As for your own writing, I already take offense at the very first sentence. Not that I took it to be false, but the sense is unclear to me. "The world is everything that is the case." The "is" is used either as a mere copula, or as the sign of equality in the fuller sense of "is the same as". While the "is" of the subordinate clause is obviously a mere copula, I can only understand the "is" of the main clause in the sense of an equality sign. Up to here I believe no doubt is possible. But is the equation to be understood as a definition? That is not so clear. Do you want to say, "I understand by 'world', everything that is the case"? Then "the world" is the explained expression, "everything that is the case" the explaining expression. In this case nothing is thereby asserted of the world or of that which is the case, but if anything is to be asserted, then it is something about the author's use of language. Whether and how far this use might concur with the language of everyday life is a separate matter, which is, however, of little concern to the philosopher once he has established his use of language.

But in an equality a thought can also be expressed, which extends our knowledge essentially, if it is acknowledged as true. Every recognition is a piece of knowledge of this kind.

Ein Planet wird z.B. als ein schon früher beobachteter wiedererkannt. Wir haben dann zunächst zwei Namen: den Namen, den der Planet schon früher erhalten hat, und zweitens den Namen, den ich ihm jetzt beigelegt habe, wenn auch nur in der Form "der eben jetzt von mir beobachtete Planet". Der Astronom bildet nun zunächst versuchsweise, etwa in Form einer Frage die Gleichung "Ist Eros der eben jetzt von mir beobachtete Planet?" Diese Frage hat für den Astronomen einen Sinn. Er ist überzeugt: sie muss entweder bejaht oder verneint werden. Der Name "Eros" hat für ihn einen Sinn und ebenfalls der Ausdruck "der eben jetzt von mir beobachtete Planet". Und jeder dieser Namen hatte diesen Sinn schon, bevor die Gleichung gebildet wurde. Keiner dieser Namen erhält seinen Sinn erst kraft dieser Gleichung wie im Falle der Definition. Auch dass jeder dieser Namen bedeutungsvoll sei, stand dem Astronomen schon fest, bevor er die Frage bildete. In diesem Falle wird keinem der Namen erst durch die Frage oder durch das Urteil, in dem sie bejaht wird, seine Bedeutung beigelegt, wie das durch eine Definitionsgleichung geschieht. Nun bejaht der Astronom—will ich einmal annehmen—die Frage. Damit setzt er nichts über seinen Sprachgebrauch fest wie im Falle der Definition; aber er gewinnt dadurch eine neue Erkenntnis, die wesentlich wertvoller ist, als eine blosse Folgerung aus dem allgemeinen Identitätsgesetze $a = a$. Wenn man durch die Definitionsgleichung $2 = 1 + 1$ dem Zeichen "2" die Bedeutung von "$1 + 1$" gegeben hat, welche ich als bekannt annehme, so gilt nun freilich $2 = 1+1$; aber es wird durch die Anerkennung dieser Gleichung eigentlich keine neue Erkenntnis gewonnen, sondern wir haben darin nur einen besonderen Fall des Identitätsgesetzes.

Wenn Sie nun den Satz "die Welt ist alles, was der Fall ist" nicht als Definitionsgleichung meinen, sondern in ihm eine wertvolle Erkenntnis kundgeben wollen, muss jeder der beiden Namen "die Welt" und "alles, was der Fall ist", schon vor der Bildung des Satzes einen Sinn haben, einen Sinn, der ihm also nicht erst kraft dieser Gleichung beigelegt wird. Ehe ich etwas weiter über die Sache schreiben kann, muss ich darüber in's Reine gekommen sein. Definitionsgleichung oder Wiedererkennungsurteil? oder gibt es noch ein Drittes?

Nach dem, was ich erfahren habe, ist es allerdings bei unseren trostlosen wirtschaftlichen Zuständen fast unmöglich, ein schweres Werk zum Druck zu bringen, wenn man nicht einen erheblichen Teil der Kosten trägt.

Eben ersehe ich noch aus einem früheren Ihrer Briefe, dass Sie im Idealismus einen tiefen wahren Kern anerkennen, ein wichtiges Gefühl, das unrichtig befriedigt wird, also wohl ein berechtigtes Bedürfnis. Welcher Art ist dies Bedürfnis?

Es wird mir lieb sein, wenn Sie durch Beantwortung meiner Fragen mir das Verständnis der Ergebnisse Ihres Denkens erleichtern.

Mit herzlichem Grusse in alter Freundschaft Ihr

G. Frege.

A planet is, e.g., recognized as a previously observed one. We have then for the time being two names: the name which the planet had received earlier, and, secondly the name which I have given to it now, if only in the form "the planet observed by me just now". The astronomer now forms, at first tentatively, perhaps in the form of a question, the equation "Is Eros the planet observed by me just now?" This question[19] has a sense for the astronomer. He is convinced: it must either be affirmed or denied. The name "Eros" has a sense for him and so has the expression "the planet observed by me just now". And each of these names had this sense already, before the equation was framed. Neither of these names obtains its sense only now, in virtue of this equation, as in the case of a definition. In addition, that each of these names should be meaningful was settled for the astronomer before he framed the question. In this case neither name would receive its meaning only through the question or through the judgment in which it is affirmed, as would happen in the case of a definitional equation. Now—I am supposing—the astronomer answers the question affirmatively. He thereby establishes nothing about his use of language, as in the case of the definition; but he wins through it a new piece of knowledge, which is considerably more valuable than a mere consequence of the general law of identity $a = a$. If through the definitional equation $2 = 1 + 1$ one had given the sign "2" the meaning of "$1 + 1$", which I consider as known, then obviously $2 = 1 + 1$ holds; but through the acknowledgment of this equation there is not really any new knowledge gained. Rather we have in it only a special case of the law of identity.

If, however, you do not mean the sentence "the world is everything that is the case" as a definitional equation, but want to set forth a valuable piece of knowledge, each of the two names "the world" and "everything that is the case" must already have a sense before the framing of the sentence, a sense which is therefore not only then given to it in virtue of this equation. Before I can write something further about this matter, I must have reached clarity about this. Definitional equation or recognition judgment? Or is there a third?

At any rate, from what I have learned, in our wretched economic conditions it is nearly impossible to publish a difficult work if one does not contribute a considerable part of the cost.

I just noticed from an earlier letter of yours that you acknowledge a deep and true core in idealism, an important feeling that is wrongly gratified, hence, a legitimate need. Of what sort is this need?

I would be glad if you would assist me in understanding the results of your thinking by answering my questions.

With kind regards in abiding friendship

G. Frege

---

[19] The phrase "this question sentence" [*Dieser Fragesatz*] was replaced by "this question" [*Diese Frage*].

Letter to Wittgenstein from Heinrich Scholz, 2. 4. 1936

Münster i. W., d. 2. April 1936

Herrn
Professor Dr. Ludwig Wittgenstein
Cambridge University England

Sehr geehrter Herr Professor!

Es ist mir nach grossen Anstrengungen endlich gelungen, den ganzen wissenschaftlichen Nachlass Freges in den Besitz meines logistischen Seminars zu bringen, aus dem er eines Tages auf die Universitäts-Bibliothek Münster übergeführt werden soll. Wir bereiten jetzt eine kleine Ausgabe der Kleinen Schriften Freges vor, in der auch das wertvollste Nachlass-Material veröffentlicht werden soll. Es sind zunächst zwei Bände geplant. Die Notgemeinschaft der Deutschen Wissenschaft wird unsere Ausgabe drucken.

Ich habe mich nun auch sehr um den wissenschaftlichen Briefwechsel Freges bemüht, und nicht ohne Erfolg. Das Wertvollste, was wir gewonnen haben, sind die Originale der Briefe, die Frege mit Russell gewechselt hat. Bertrand Russell hat uns diese Briefe durch eine hochherzige Schenkung zugewendet, nachdem ich ihn auf dem Pariser Kongress im September des vergangenen Jahres persönlich habe kennenlernen dürfen. Wir werden den Briefwechsel Russell-Frege in unserer Ausgabe abdrucken.

Nun geht aus den Nachlass-Papieren hervor, dass auch zwischen Ihnen und Frege Briefe gewechselt worden sein müssen, im Zusammenhang mit einer Begegnung, die Sie mit Frege gehabt haben. Ich werde annehmen dürfen, dass Sie diese Briefe noch besitzen, und würde Ihnen sehr dankbar sein, wenn Sie sich entschliessen könnten, uns diese Briefe für das Frege-Archiv zuzuwenden, in welchem alles gesammelt werden soll, was von Frege überhaupt noch erreichbar ist. Es versteht sich, dass wir gern bereit sein werden, Ihnen genaue Abschriften dieser Briefe zuzuwenden, wenn Sie Wert darauf legen.

Ich muss noch eine Bitte hinzufügen dürfen. Frege muss einige sehr wertvolle Briefe an seinen früh verstorbenen Interpreten, den ausgezeichneten Philip E. B. Jourdain, geschrieben haben. Da Sie mit Herrn Jourdain befreundet gewesen sind, so sind Sie vielleicht in der Lage, mir die gegenwärtige Addresse von Frau Jourdain anzugeben, damit ich mich an sie wenden kann. Ich würde Ihnen auch hierfür sehr dankbar sein.

Letter from Heinrich Scholz to Ludwig Wittgenstein, 2.4.1936

Münster in Westfalia, 2 April 1936

To Professor Dr. Ludwig Wittgenstein
Cambridge University, England

Dear Professor,

After considerable effort I have finally succeeded in bringing into the possession of my logical seminar the complete scientific *Nachlass* of Frege, which will be handed over to the University Library at Münster one day. We are now preparing a small edition of the short papers of Frege, in which we shall also publish the most valuable *Nachlass* material. Initially two volumes are planned. The Emergency Association of German Science will print our edition.

I have also made a real effort with regard to the scientific correspondence of Frege, and not without success. The most valuable of what we have obtained are the original letters that Frege exchanged with Russell. Bertrand Russell sent us these letters, making us an especially handsome gift, after I had met him personally in September of last year at the Paris Congress. We will publish the Russell-Frege correspondence in our edition.

Now it seems from the *Nachlass* papers that there must have been an exchange of letters between you and Frege, in connection with a meeting that you had with Frege. I would presume that you still possess these letters, and would be very thankful if you would decide to send them to us for the Frege archive, where everything will be collected of Frege's that is still available. Obviously we would be glad to send back to you accurate transcriptions of the letters if you wished.

I should like to make one more request. Frege must have written some very valuable letters to the excellent Philip E.B. Jourdain, the interpreter who died at a young age. Because you were a friend of Mr. Jourdain, perhaps you would be in a position to give me the present address of Mrs. Jourdain, in order that I might write to her. I would be very grateful for this too.

Schliesslich möchte ich Ihnen noch sagen dürfen, dass ich aus der Zeit, in der wir gemeinsam in Kiel gewirkt haben, ein guter alter Freund von Herrn Schlick bin, und dass ich viele Briefe von ihm habe, die ganz wesentlich angefüllt sind mit dem, was er Ihnen schuldig geworden ist.

Ich bitte Sie, dass Sie mich Herrn Braithwaite empfehlen.

In grösster Hochschätzung

Ihr sehr ergebener

Heinrich Scholz

o. Prof. d. Philosophie u. Logistik a. d. Universität
Münster i. Westf.

In closing, allow me to say that I was a close and old friend of Mr. Schlick's when we worked together in Kiel, and that I have many letters from him that are filled throughout with what he owed essentially to you.

Please pass on my greetings to Mr. Braithwaite.

In great respect

Your devoted

H. Scholz
Prof. of Philosophy & Logistic
At the University of Münster in Westfalia

Letter from Wittgenstein to Heinrich Scholz, 9. 4. 1936

Wien. 9.4.36.
Ständige Adresse: Trinity College
Cambridge

Sehr geehrter Herr Professor!

Ich bestätige dankend den Empfang Ihres Schreibens vom 2.4. Ich besitze zwar einige wenige Karten & Briefe Freges, sie sind aber rein persönlichen, nicht philosophischen, Inhalts. Für eine Sammlung der Schriften Freges haben sie keinerlei Wert; wohl aber für mich einen Erinnerungswert. Der Gedanke widerstrebt mir, sie einer öffentlichen Sammlung zur Verfügung zu stellen.

Herr Jourdain war, als ich ihn kannte, meines Wissens, nicht verheiratet. Ich werde mich aber noch genauer erkundigen & Ihnen dann berichten.

Hochachtungsvoll

Ludwig Wittgenstein

Letter from Wittgenstein to Heinrich Scholz, 9.4.1936

Vienna, 9. 4. 1936
Permanent address: Trinity College
Cambridge

Dear Professor,

Thank you for your letter of 2.4. I possess only a few cards and letters of Frege's whose contents are, however, purely personal and not philosophical. For a collection of Frege's writings they are of no value whatsoever, although they have a sentimental value for me. The thought of setting them up in a public collection perturbs me.

When I knew him Mr. Jourdain was to my knowledge unmarried. I shall, however, better inform myself and then report to you.

Respectfully

Ludwig Wittgenstein

The Frege-Wittgenstein Correspondence: Interpretive Themes

Juliet Floyd

Twenty-one cards and letters from Frege to Wittgenstein—the totality of the correspondence between them presently known to exist—were discovered in 1988, long after elaborate and far-reaching interpretive traditions had grown up around each philosopher.[1] It is unlikely that these missives will of themselves radically reshape our understanding of either. But for historians of logic and analytic philosophy, as well as for anyone interested in German and Austrian intellectual history at the time of the First World War—and especially Wittgenstein's and Frege's places within it—these are significant and interesting documents.

First and foremost, the cards and letters are accessible and engaging reading in their own right, documenting in a concrete way the course of intellectual exchange between two great philosophers, as well as some of Frege's own wartime observations of life in Germany. Second, they make a bit more vivid the nature of the relationship between Frege and Wittgenstein, a relation that unfolded over 9 years during a period that was crucially formative in Wittgenstein's early development, and hence in the development of early twentieth century philosophy as a whole. Third, the letters provide a new kind of textual factor shaping reflection on the overall significance and nature of Frege's philosophical impact on Wittgenstein, and vice versa. For they contain a record of Frege's highly critical reactions to the *Tractatus* manuscript, which Wittgenstein had sent to him in December 1918 after having had the

[1] The letters from Frege to Wittgenstein were first published in an issue of *Grazer Philosophische Studien* as "Gottlob Frege: Briefe an Ludwig Wittgenstein", eds. A. Janik and P. Berger, in vol. 33/34, *Wittgenstein in Focus — Im Brennpunkt: Wittgenstein*, eds. Brian McGuinness and Rudolf Haller (Amsterdam: Rodopi, 1989), pp. 5–33, and again, with editorial revisions and commentary, in the CD-ROM of Wittgenstein's complete known correspondence distributed by Intelex, *Ludwig Wittgenstein: Briefwechsel* (Innsbrucker elektronische Ausgabe 2004), eds. Monika Seekircher, Brian McGuinness and Anton Unterkircher. They are translated in this volume; see the preface to this translation for editorial commentary on their history.

E. De Pellegrin (ed.), *Interactive Wittgenstein*, Synthese Library 349,
DOI 10.1007/978-1-4020-9909-0_3,

manuscript rejected by the literary publisher Jahoda and Siegel.[2] And they also contain his reaction to Wittgenstein's frank criticisms (now lost, with Wittgenstein's side of the correspondence) of Frege's later highly influential philosophical essay "Der Gedanke" ("Thoughts"), an essay that, as the letters also establish, Frege sent to Wittgenstein in an offprint.[3]

What immediately strikes a reader of this correspondence is its tone of personal and intellectual closeness; a tone unique within Frege's published academic correspondence and something of a surprise for Wittgenstein scholars, who may not have known of the extent of this dimension of their relationship until the letters were published. Clearly this was a singular meeting of souls who shared mutual respect for one another's intellectual tenacity and sensibility, hope for collaboration, and philosophical values and interests (in clarity and intellectual honesty, in the importance of the new mathematical logic, in the nature and importance of logic to philosophy). The writings culminate, in spite of this closeness, in unanswered criticisms and an end to philosophical discussion and/or any imagined collaboration. Scholars previously knew of this result from remarks made, not only by Wittgenstein's sister Hermine, but also by Wittgenstein himself, in letters to Russell and Ficker and later remarks to Geach.[4] But here one may read the closing gesture in explicit form, as written down by Frege.

How are we to weigh the letters against the backdrop of recent discussion—wide-ranging and increasingly voluminous—about how to understand the

[2] Frege received the manuscript via Wittgenstein's sister Hermine in late 1918 or early 1919, but did not reply until 28 June 1919; see the Chronology in my Preface to the translation, as well as von Wright, "The Origin of the *Tractatus*", p. 76 and related correspondence in Ludwig Wittgenstein, *Cambridge Letters: Correspondence with Russell, Keynes, Moore, Ramsey and Sraffa*, eds. B. McGuinness and G.H. von Wright (Cambridge, MA: Blackwell, 1995). See also footnote 31 below.

[3] See Frege to Wittgenstein of 12 September and 15 October 1918, and 3 April 1920.

[4] G.H. von Wright analyzed this correspondence in detail before the discovery of the Frege letters in "The Origin of the Tractatus" (in his *Wittgenstein. With letters from Wittgenstein* (Oxford: Blackwell, 1982/Minneapolis, MN: University of Minnesota Press, 1983) and also reprinted on the CD-ROM *Ludwig Wittgenstein: Briefwechsel*). This essay remains essential reading for those interested in the origins and composition of the *Tractatus*. So too are the introduction to B. McGuinness and J. Schulte, eds., *Logisch-philosophische Abhandlung/Tractatus Logico-Philosophicus, Kritische Edition* (Frankfurt, Suhrkamp Verlag, 1989) and essays touching upon this topic in Brian McGuinness's *Approaches to Wittgenstein: Collected Papers* (New York: Routledge, 2002). For Hermine's comments on the relationship with Frege, see her "My Brother Ludwig", in *Recollections of Wittgenstein*, ed. Rush Rhees (New York: Oxford University Press, revised edition 1984), pp. 1–11, especially pp. 5–6. For Geach's anecdote, see the Preface to Frege, *Logical Investigations*, ed. and trans. P.T. Geach (New Haven, CT: Yale University Press, 1977).

philosophical relations between Frege and Wittgenstein? Largely on the basis of the letters, Frege's biographer Lothar Kreiser has written that in the face of the *Tractatus* and Wittgenstein's efforts to explain it, both Frege and Wittgenstein simply "gave up" trying to understand each other.[5] This is surely not true of Wittgenstein, who, as is well known, returned repeatedly throughout his subsequent philosophical life to consideration of Frege's writings and turns of phrase and thought, as well as the content of their conversations.[6] But Kreiser's point may have been true for Frege. "It would remain a riddle" to Frege, Kreiser writes, "in what his influence on L. Wittgenstein might really have consisted, and for what reason he was thanked in the Preface to the *Tractatus*".[7] So far as we know, the friendship and correspondence between them was not further pursued by either after 1920 (Frege was to die in 1925).[8] Whether from Frege's side this had to do primarily with his retirement and lack of energy, or his philosophical reservations about the *Tractatus* and/or Wittgenstein's negative reactions to "Der Gedanke" we shall never know.

In any case Kreiser's comments lead us naturally to the question whether readers ought to classify the correspondence as reflecting nothing more than a biographical curiosity of little interest to philosophy, an exchange between two thinkers that went nowhere.

In his 1989 editor's foreword to the initial publication of the correspondence, Allan Janik departed from this view, suggesting that the depth of differences between Wittgenstein and Frege—evinced especially in Frege's critical remarks about the *Tractatus*—indicate something important about very different conceptions of clarity informing these two founding figures of early analytic philosophy. As Janik wrote,

> Frege's letters about the "Tractatus" convey not only the respect and friendship he felt for Wittgenstein, but also the two thinkers' utterly distinct conceptions of clarity (*Klarheit*)—a theme which continues to demand the attention of philosophers if we are to grasp the deepest distinctions separating one champion of an analytical philosophy from another.[9]

Janik does not specify the differences he sees at work between the "utterly distinct" conceptions of clarity informing Frege's and Wittgenstein's philosophies, but since the goal of conceptual clarity lies at the heart of the ana-

[5] Lothar Kreiser, *Gottlob Frege Leben-Werk-Zeit* (Hamburg: Felix Meiner Verlag, 2001), p. 580.

[6] Reck, "Wittgenstein's 'Great Debt' to Frege", in Reck, ed., *From Frege to Wittgenstein* (New York: Oxford University Press, 2002), pp. 3–38 summarizes the biographical data and contains a discussion of the Frege-Wittgenstein correspondence, as well as a few tentative suggestions about how we ought to be viewing the question of Frege's influence on Wittgenstein.

[7] Lothar Kreiser, *Gottlob Frege Leben-Werk-Zeit*, p. 580.

[8] Compare Hermine Wittgenstein, "My Brother Ludwig", pp. 5–6.

[9] Introduction to "Gottlob Frege: Briefe an Ludwig Wittgenstein", p. 7.

lytic tradition's self conception, his remarks claim for the Frege-Wittgenstein correspondence a special place within our understanding of the tradition's early development. If Janik is right, a fundamental and important philosophical break already existed at the origins. Since making this remark, Janik has gone on to examine the influence of Frege on Wittgenstein in more philosophical detail, though not primarily with an eye on the Frege-Wittgenstein correspondence. He of course does not deny that Frege had an impact upon Wittgenstein—noting that Wittgenstein himself placed Frege on the list of those who had most influenced him.[10] In particular, Janik stresses, Frege's anti-psychologism and style left their mark upon Wittgenstein, along with the theme of breaking the hold of misguided philosophical views of word-thing meaning relations by an appeal to contextualism.[11]

Of course, the content, basis, and implications of the anti-psychologism and contextualism have been the subject of much discussion, both about Frege's and Wittgenstein's philosophies. When we raise the question of the relationship between Frege and Wittgenstein, we are thus on the brink of larger, profound questions about gating ideas in early analytic philosophy and our relationship to them. How far did Frege and Wittgenstein really manage to work themselves into each others' point of view? Apart from Frege's style and intellectual tenacity and purity, which certainly left their marks on Wittgenstein,[12] is Frege's influence on Wittgenstein best seen as that of a thinker who posed problems that stimulated Wittgenstein, or instead as someone whose basic ideas were taken over by Wittgenstein, and perhaps thought through to a more thoroughgoing conclusion?[13] Was Wittgenstein's development largely independent of Frege, overlapping where the limitations of alternative approaches seemed most clear?[14] How much philosophical agreement underlay their disagreements? At which time? On which issues and grounds? What relevance do their answers have to contemporary philosophical discussion of their views?

It is clear that the letters alone cannot secure an interpretation of the Frege-Wittgenstein relation; we do best, in considering texts relevant to understanding this—both in matters of philosophical substance and in answering questions of influence and development—to look to a wide range of texts and the philosophical issues themselves, and avoid viewing the letters as an interpretive silver bullet. It seems unlikely, in fact, that answers will be forthcoming

[10] See the 1 April 1932 list of figures who Wittgenstein said had most influenced him, at item 154, 16r in his *Nachlass.*

[11] *Assembling Reminders: Studies in the Genesis of Wittgenstein's Concept of Philosophy* (Stockholm: Santérus Press, 2006).

[12] Cora Diamond, "Inheriting Frege: The Work of reception, as Wittgenstein did it", forthcoming in *The Cambridge Companion to Wittgenstein.*

[13] Those in this tradition include Geach, Diamond, Hintikka, and Ricketts.

[14] For this view see Goldfarb, "Wittgenstein's Understanding of Frege: The Pre-Tractarian Evidence", in E. Reck, ed., *From Frege to Wittgenstein* (New York: Oxford University Press, 2002), pp. 185–200.

from scrutiny of any smallish portion of the textual evidence alone—though such scrutiny is essential, of course, in arranging what evidence exists. To a large extent, we understand the letters by looking at surrounding texts.

This does not imply, however, that the letters have no philosophical significance whatsoever. Few interpreters of Wittgenstein and Frege have attempted to discuss the extent to which their contents shed unique light on such interpretive philosophical matters. And the biographies that have so far been written on both Frege and on Wittgenstein, while excellent, have also failed to address them within the larger context of a narrative about the origins of early analytic philosophy as a whole.[15]

While philosophy is not reducible to biography or vice versa, I also do not think it either possible or desirable wholly to abstract the life or historical context in which a philosopher writes from an interpretation of the significance of his or her writings.[16] In the case of a philosophical and personal correspondence this is especially important to bear in mind. To set the letters into proper light we must emphasize, not only philosophical themes and problems raised by the correspondence, but also certain contingencies of the historical situation in which the correspondents found themselves where these may be useful for assessing the philosophical significance of the letters. In what follows I shall be standing very much on the shoulders of Wittgenstein's biographers, Brian McGuinness and Ray Monk, and relying on the earlier, ground-breaking scholarly work (pursued before the discovery of the correspondence) of G.H. von Wright. My aim is not to give crucial philosophical weight to the letters, but to canvas several points surrounding their contents. I shall highlight primarily the biographical context (in Section I) and then (in Section II) some of the more philosophical issues.

---

[15] Brian McGuinness's biography of Wittgenstein's early life, *Wittgenstein: A Life, Young Ludwig 1889–1921* (Berkeley, CA: University of California Press, 1988) was published before the discovery of the letters; Ray Monk's *Ludwig Wittgenstein: The Duty of Genius* (New York: Free Press, 1990) (especially at pp. 151ff., 174ff.) and Lothar Kreiser's *Gottlob Frege Leben-Werk-Zeit* were published afterwards, and do weave references to the letters into the discussion of their subjects, though without emphasizing the questions I am raising here.

[16] On the theme of biography and philosophy, see my review of J. Klagge, ed., *Wittgenstein: Biography and Philosophy, Notre Dame Philosophical Reviews* (2002.06.04) at http://ndpr.icaap.org/content/current/floyd-klagge.html. On the broader question of the historical contextualization of analytic philosophy, see my introduction, with S. Shieh, to J. Floyd and S. Shieh eds., *Future Pasts: Perspectives on the Analytic Tradition in Twentieth Century Philosophy* (New York: Oxford University Press, 2001).

I

Wittgenstein was the first to answer our question as to what philosophical significance, if any, the correspondence contains. For Heinrich Scholz explicitly wrote to him about the letters (2 April 1936).[17] Scholz had made it clear that he had evidence of the existence of a correspondence between Frege and Wittgenstein "in connection with a meeting that you [Wittgenstein] had with Frege".[18] He explained that his aim was to publish a collection of Frege's "scientific correspondence" and to create a Frege archive at the University of Münster. Then, with the perfectly appropriate but distinctive tone of a seeker of donations, Scholz cited Russell's "handsome" gesture in donating the originals of his correspondence with Frege to the archives (originals which included, we may presume, their remarkable exchanges about Russell's discovery of his paradox in 1902),[19] and urged Wittgenstein to follow suit. He was proposing, in other words, not only to read the contents of the Frege-Wittgenstein letters with an eye toward their publication, but also to retain the originals for posterity within the Frege Archive. He then asked for Wittgenstein's help in contacting Phillip Jourdain's widow, in case such a person existed, to obtain further Frege letters.[20] Finally, in closing, Scholz took up the role of an appreciator of Wittgenstein's work, adducing Schlick as a mutual close acquaintance and stating that the "many" letters he possessed from Schlick were "filled throughout" with what Scholz believed Schlick to have "owed essentially", philosophically speaking, to Wittgenstein.

Wittgenstein wrote back to Scholz within a week, that is to say, fairly rapidly. He said he was under the impression that Jourdain had been unmar-

---

[17] I have included the Scholz-Wittgenstein exchange of letters from 1936 in the translation in this volume.

[18] Reference to a record of this meeting is contained in Scholz List 2, now in the Scholz Archiv at Münster (see my Preface to the translations, in this volume, for citations to this list).

[19] The 1902 exchange between Frege and Russell is translated in Jean van Heijenoort, ed., *From Frege to Gödel: A Sourcebook in Mathematical Logic, 1879–1931* (Cambridge, MA: Harvard University Press, 1967), pp. 126–128, along with a stirring letter by Russell to van Heijenoort praising Frege's intellectual honesty, dedication, and integrity. (Van Heijenoort evidently worked with copies of the original letters.)

[20] This was presumably because Scholz knew of the March 29, 1913 letter from Jourdain to Frege in which Jourdain says that he and Wittgenstein "were rather disturbed" by the idea that Frege might be writing a third volume of the *Grundgesetze*, and suggest a translation of earlier parts of the book into English instead. Frege approved the project in his reply (cf. Frege, *Philosophical and Mathematical Correspondence*, eds. G. Gabriel et al. (Chicago: University of Chicago Press, 1980). As Reck notes ("Wittgenstein's 'Great Debt' to Frege", p. 12), this indicates, minimally, that Wittgenstein was interested enough in Frege's work to contribute to its translation, and that Frege trusted Wittgenstein enough to approve of his involvement in this venture. (This translation project was not completed.)

ried, but would "better inform" himself and write back if there were more to say (it seems he never did). Scholz's remarks about Schlick's letters owing so much to his influence were unlikely to have impressed Wittgenstein favorably, and he did not reply to these at all. Wittgenstein had already written to Schlick years earlier urging him to "tone down the fanfare stuff" in publicly praising Wittgenstein's *Tractatus*, because "for 1,000 reasons it was no triumph".[21] This was a reaction to Schlick's essay "The Turning Point in Philosophy", which Schlick had sent to him when it appeared in 1930.[22] Schlick had explicitly placed the *Tractatus* on a world-historical stage, writing that, even in relation to Leibniz, Russell and Frege, Wittgenstein was "the first to have pushed forward to the decisive turning point" in philosophy. Disturbed by the growing tendency to affiliate his early work with the Vienna Circle as a movement, realizing through his discussions with Ramsey the technical limitations of his early work, Wittgenstein was clearly worried that Schlick's hyperbolic praise of him bordered on the ridiculous. In reaction, he reminded Schlick of the saying from Nestroy that would later become the motto of *Philosophical Investigations*: "do not forget that handsome saying of Nestroy's ... that progress has this in it, that it always looks greater than it is."[23]

As for Frege's letters to him, Wittgenstein acknowledged to Scholz that they were in his possession (although, as we may plausibly assume, they were being held by or for him in Vienna, not in Cambridge).[24] In refusing Scholz's request for access to the letters, Wittgenstein cited three reasons.

(1) The cards and letters are few in number and their contents are "purely personal and not philosophical", having "no value whatsoever" for a collection of Frege's [scientific] writings;
(2) The cards and letters have a "sentimental" value for Wittgenstein;
(3) Wittgenstein is "perturbed" by the idea of setting them up in a public collection of Frege's work.

---

[21] Wittgenstein to Schlick (18 September 1930); see *Briefwechsel.*

[22] It appeared in the first number of *Erkenntnis* vol. I (in 1930/1931): 4–11; for Schlick's paper in English see Ayer, ed. *Logical Positivism* (Glencoe, IL: The Free Press, 1959), pp. 53–59, especially p. 54.

[23] For more on the motto and its meaning, see David Stern, Wittgenstein's *Philosophical Investigations: An Introduction* (Cambridge, 2004) and my "Homage to Vienna: Feyerabend on Wittgenstein (and Austin and Quine)", in *Paul Feyerabend (1924–1994): Ein Philosoph aus Wien*, eds. K.R. Fischer and F. Stadler, *Veröffentlichungen des Instituts Wiener Kreis*, vol. 14 (Springer Verlag, 2006).

[24] It seems plausible to assume that the letters were being held for Wittgenstein alongside the other pieces of correspondence with which they were later discovered, by his arrangement or perhaps that of a member of the family acting as his representative. This particular collection of over 500 letters was large, and it seems unlikely Wittgenstein would have had it shipped to Cambridge with him. Because the circumstances surrounding the later discovery of the correspondence are so murky, however, we know next to nothing of the history of this collection of letters.

In rejecting Scholz's eminently reasonable appeal for scholarly help, Wittgenstein was, it seems, not only needlessly dismissive of Scholz and the Frege archive project, but also positively dishonest with Scholz about the contents and philosophical significance of the correspondence. The overall impression, at least initially, is of a selfishly highhanded and impetuous man, unconcerned with scholarship, protective of his own vanity and reputation, unwilling to take any time to help a fellow researcher, and dismissive of Frege's philosophical remarks.

As I see matters, however, this initial impression is not all there is to say about the Scholz–Wittgenstein exchange. Even if the reasons Wittgenstein gave to Scholz constituted but a part of the truth, each contained large grains of it.

It should of course be asked whether Wittgenstein's decision to reply to Scholz as he did was nothing more than a selfish outburst by a philosopher who deemed the academic study of anything important impossible. McGuinness has raised the issue explicitly concerning the early Wittgenstein, writing that "Ludwig's own inclinations", at least in the period around 1919, were hostile to any form of study, and that "the idea of academic study of anything important is explicitly rejected in his book [the *Tractatus*]", at least as a life choice for Ludwig at that time, if not as a matter of philosophical principle.[25] If one grants that such an attitude was in place in 1919, at issue is the question whether a sufficiently strong residue of it extended into the 1930s in such a way as to explain, on its own, Wittgenstein's reply to Scholz.

Here I would answer in the negative. For I take Wittgenstein to have been acting in what is an understandable and rationally calculated way, attempting to do what he took to be the appropriate thing to protect the interests of all concerned—including, of course, his own. This is not to deny that Scholz, a working logician and founder of an important academic archive, had a right to feel that he had not been treated as well as he might have been, or that Wittgenstein was never fully devoted to academic professionalism. Nor is it to deny that Wittgenstein's temperament, including what he himself was repeatedly to call his own "vanity" in the prefaces and forewords to his projected books, played no part, either in his refusal to divulge the contents of the letters to Scholz or in his earlier behavior, intellectual and personal, with Frege.[26] But it is to suggest that we ought to assess Wittgenstein's decision in context, and allow ourselves to entertain the interesting question whether the letters (both the significance of their contents and their archival location) properly belong, ultimately, within the context of Wittgenstein's life's corpus rather

[25] *Wittgenstein: A Life, Young Ludwig 1889–1921*, p. 284.

[26] Remarks concerning the dangers both of vanity and of false humility in putting a philosophical work before the public find their way into the Preface to the *Tractatus* implicitly, but are made explicit in the Foreword to *Philosophical Remarks* and the Preface to *Philosophical Investigations.*

than with Frege's. It seems to me that Wittgenstein's 1936 decision, at the very least, correctly attached them to the former and not the latter.

Wittgenstein's 1936 reply to Scholz expressed a complex desire on his part to achieve a number of differing goals. Knowing what we now do about his preoccupations and state of mind in 1936, we can consider the forces and questions in play for him at that time.

The first reason Wittgenstein gave for not sharing the letters with Scholz is that the cards and letters are "purely personal and not philosophical", having "no value whatsoever" "for a collection of Frege's [scientific] writings". With the latter point it is difficult to disagree: the Frege-Wittgenstein correspondence is not nearly of the same importance to an understanding of Frege's development as a logician and philosopher as are, for example, his correspondence with Husserl, with Russell, with Hilbert and with Peano, of which scholars have rightly made a great deal. No fundamental points of symbolic logic or mathematics are touched on in the exchanges. And while Frege's philosophical ideas—above all about sense (*Sinn*) and reference (*Bedeutung*)—play an explicit role in his criticisms of the *Tractatus*, there are no new twists to the central lines of Frege's thought revealed here. Wittgenstein was often to refer to Frege in subsequent writings, but singled out other issues to criticize than those broached in the letters (he focuses mostly on Frege's criticisms of formalism, his definition of number, his view of logic as a maximally general science, his view of concepts, thought, and of the privacy of psychological images and sensations). Frege writes to Wittgenstein explicitly that he feels that even his essay "Der Gedanke" has "perhaps little new in it; but perhaps said in a new way and therefore more intelligible to some" (Frege to Wittgenstein 12 September 1918). Even if that essay's importance is by now historically confirmed, there is arguably little direct light shed on it by consideration of Frege's letters to Wittgenstein, including Wittgenstein's responses to Frege as indicated in his replies.

Nevertheless Wittgenstein's claim, that the cards and letters are "not philosophical", is obviously misleading. The criticisms Frege makes of the *Tractatus*, are explicit, fairly detailed, and harsh. During the war years there was a complete cessation of philosophical exchange between Frege and Wittgenstein. What we know of their philosophical conversations before the war is provided by what scholars have had in hand for some time, namely, the Scholz lists and related correspondence and testimony of Wittgenstein and others. So what we learn of their exchanges after the war is given by the final four letters of the Frege-Wittgenstein correspondence alone (June 1919–April 1920). It is striking that this final chapter in their recorded exchanges shows so vividly Frege's inability to appreciate the *Tractatus*, his suggestions that Wittgenstein revise the manuscript, and his highly tentative willingness to aid in securing its publication (and not in the form Wittgenstein sent it to him).

The letters show that Frege was not able to get far with either the manuscript or Wittgenstein's letters to him explaining it. As we know, Wittgenstein felt at the time he received Frege's reactions that they were useless; he wrote to Russell in 1919 that Frege had not "understood a single word" of the *Tractatus*, that it was "VERY hard not to be understood by a single soul!", and that he was "thoroughly exhausted" by his efforts to give "simple explanations" to Frege.[27] Clearly in 1919 he honestly felt that Frege's criticisms of the manuscript were of no philosophical worth.

There is, however, an interesting question whether he felt differently about this in 1936, after his own thinking had evolved beyond the *Tractatus*. It is worth noting that he was later to propose that the *Tractatus* be published beside the *Investigations*, to show his later thought in its appropriate light.[28] This could be taken to suggest that he still did not take to heart any of the suggestions for improvement and rewriting that Frege had urged. Nevertheless, evidence does exist that the philosophical points discussed in their correspondence remained with him long afterwards, as we shall see (in Section II below).

What of Wittgenstein's remark to Scholz that the cards and letters are merely "personal" in character? This is true of all but the final four letters—indeed, this is what makes the correspondence so fascinating to read. We see Frege make remarks about his neighbors, about jokes in the local newspapers, about the deaths of relatives. We even gain what may be some small further insight, through his remarks on the wartime campaigns, of his thinking about Germany's place in the war.[29] For Frege, Wittgenstein was a young soldier to be respected and supported for his sacrifices on the battlefield, as well as a gifted student of (Frege's and Russell's) logic 41 years his junior. Like Russell, Frege had seen in Wittgenstein a bright young hope for the future of logic, a gifted interlocutor (indeed, possible collaborator) willing and able to

[27] Wittgenstein to Russell, 19 August 1919, 6 October 1919; see *Briefwechsel* and *Cambridge Letters*.

[28] Item 128, p. 51, from 1943, in the *Nachlass*.

[29] This is not the place to discuss Frege's political views, which have been treated elsewhere by Kienzler, Kreiser, and Uwe-Dathe (see my footnote 2 to the translation of Frege's 2 August 1916 letter in this volume). But an example of the kind of remark I have in mind (noted by Burton Dreben) is contained in Frege's card to Wittgenstein of 28 August 1916, where Frege mentions with great trepidation the entry of Romania into the war. While Frege's nervousness about this may be partly intended to express concern for Wittgenstein, who is fighting on the Eastern front, Frege fails in his letter to Wittgenstein of 26 April 1917 even to mention the entry of the United States into the war (on 6 April 1917), alluding instead to the successes of the U-Boat campaign in the Atlantic. Was this an underestimation (perhaps typical in Germany at the time) of the overwhelming role that was to be played by the emerging North American industrial power in the subsequent months of the war, or was it part of an effort to encourage Wittgenstein in the face of worrying news? Compare Monk's remarks in *Ludwig Wittgenstein: The Duty of Genius*, p. 151.

discuss his logical doctrines with him.[30] Unlike Russell, when Wittgenstein's enigmatic manuscript came to him, he did not make sense of it, and it clearly disappointed him. This may partly explain the long delay in his responding to Wittgenstein's repeated requests for judgment on it. It is, moreover, worth remembering that the letters and cards were written by an aging logician primarily concerned about the lack of academic and intellectual recognition of his work and about the political future of Germany, and entangled in arranging life in his retirement during the war years.[31] Frege's health was not steady, as he attests more than once in the correspondence. Even before he received Wittgenstein's manuscript he declined invitations to visit him in Vienna and complained of his lack of strength.[32]

There is, however, above and beyond all these factors, another dimension to the "personal" side of the correspondence that must be mentioned. The letters document that Wittgenstein provided Frege with a substantial sum of money in the early part of 1918, the very year that he was to bring the manuscript of the *Tractatus* to its final form, writing in the Preface of his primary debt to "Frege's great works", and then making a series of strenuous and ill-fated efforts—including appeals to Frege—to get his manuscript published.[33]

At the time he arranged for the gift to Frege, Wittgenstein very likely viewed his act of financial beneficence—which fell squarely within his wealthy family's and his own (pre-1918) tradition of sponsoring intellectuals and artists[34]—as a tribute to Frege's logical work, as well as an alleviation of

---

[30] Kreiser (*Gottlob Frege Leben-Werk-Zeit*, p. 577) writes that Wittgenstein's visit to Frege in 1911 was "a great encouragement" to Frege. Compare Frege's letters to Wittgenstein of 1918.

[31] Lothar Kreiser canvasses possible connections between the delay in Frege's reply to Wittgenstein after receiving the manuscript of the *Tractatus* and the practicalities of Frege's life both in his biography of Frege (*Gottlob Frege Leben-Werk-Zeit*) and in "Alfred", in G. Gabriel and W. Kienzler eds., *Frege in Jena: Beiträge zur Spurensicherung* (Würzburg: Königshausen & Neumann GmbH, 1997), pp. 68–83.

[32] See Frege to Wittgenstein 12 April 1916, 2 August 1916, 28 June 1919. Kreiser discusses Frege's weak nerves and at times fragile condition in *Gottlob Frege Leben-Werk-Zeit*, pp. 513ff.

[33] See the letter from Frege to Wittgenstein of 9 April 1918, translated in this volume and in German on the CD-ROM *Ludwig Wittgenstein: Briefwechsel*. The foreword was found at the end of the manuscript that has come to be known as the *Prototractatus* (MS 104 in von Wright's catalog). For discussion of its status, see von Wright, "The Wittgenstein Papers" and "The Origin of the *Tractatus*", both in his *Wittgenstein*. Compare McGuinness and Schulte's introduction to their edition *Ludwig Wittgenstein: Logische-philosophische Abhandlung/Tractatus Logico-philosophicus, Kritische Edition*. On the gift's significance for Frege's financial situation, see Kreiser, *Gottlob Frege Leben-Werk-Zeit*, pp. 497, 505–5, 569.

[34] Among others whom Wittgenstein supported (albeit anonymously, through Ficker) were Karl Kraus, and the poets Rilke and Trakl. Their correspondence with him (after learning of his support) were discovered alongside the Frege-

what he perceived to be Frege's genuine financial need.[35] But the following year, when Wittgenstein sought Frege's help in publishing his manuscript, this act of beneficence would run the risk of raising a more mixed or heightened interpretation of motives, at least in his own mind.[36] For Wittgenstein wanted from Frege not only honest intellectual judgment of his work, but also advice and support in bringing it before the world as a publication. Frege for his part certainly responded to these requests with full intellectual honesty, even if not with wholehearted enthusiasm: he stated that he would be willing to write to the editor Professor Bauch only "that I have come to know you as a thinker to be taken rather seriously", and not about "the treatise itself", for about this "I can render no judgment, not because I am not in agreement with the content, but because the content is not sufficiently clear to me".[37]

This brings us to the second reason Wittgenstein offered to Scholz, that the cards and letters had "a sentimental value" for him. Wittgenstein cannot have forgotten the pain Frege's disappointment in the *Tractatus* had caused him 17

Wittgenstein letters, and might therefore usefully be compared with Frege's to him. (They are on the CD-ROM *Ludwig Wittgenstein: Briefwechsel* (Innsbrucker elektronische Ausgabe 2004).) Note that support of intellectuals and artists was not the only kind of charitable giving in which Wittgenstein engaged during this period of his life. McGuinness reports that according to Wittgenstein's sister Hermine, around late 1916 or 1917 Ludwig gave 100,000 crowns for the purchase of better howitzer guns for the front (*Wittgenstein: A Life*, p. 257)—the gift of a soldier and an engineer, not merely an artist or humanitarian. Compare Monk, *Ludwig Wittgenstein: The Duty of Genius*, pp. 106ff.

[35] Lothar Kreiser has said that without Wittgenstein's gift Frege could not possibly have purchased a house and retired in his home town of Bad Kleinen, Mecklenburg; moreover, without that gift, by the end of the First World War Frege would have been living "on the threshold of poverty" (*Gottlob Frege Leben-Werk-Zeit*, p. 566). Peter Geach's report of Wittgenstein's remarks about an early visit to Frege, in which Wittgenstein says he had heard that Frege was very poor (G.E.M. Anscombe and P. Geach, *Three Philosophers* (Oxford: Basil Blackwell, 1961), pp. 129–130, though relevant to the question of perceptions, may reflect Wittgenstein's own privileged upbringing and youthful dandyism more than it does Frege's actual financial situation in 1913. Compare the follow-up correspondence between Geach and Frege's biographer Kreiser, quoted in Kreiser's *Gottlob Frege Leben-Werk-Zeit*, p. 498.

[36] On the topic of mixed motives in such acts of financial subvention of intellectuals, compare Monk, *Ludwig Wittgenstein: The Duty of Genius*, p. 108:

> It is difficult to resist the conclusion that Wittgenstein's [1914] offer to Ficker [of 100,000 crowns] was motivated not only by philanthropy, but also by a desire to establish some contact with the intellectual life of Austria. After all, [in 1914] he had severed communication with his Cambridge friends, Russell and Moore, despairing of their ever understanding his ideals and sensitivities. Perhaps among Austrians he might be better understood.

[37] See Frege to Wittgenstein of 30 September 1919, translated below.

years before, when he felt most committed to trying to publish his manuscript and at the same time most devastated by the effects of the war,[38] squeamish and vulnerable about the extent of his own pride and vanity in attempting—through several rejections by well-known publishers—to place the book before the public.[39] The whole event was embarrassing and traumatic. As G.H. von Wright has written of what he called "the long and troubled history of the publication of the *Tractatus*", "it is obvious that Wittgenstein was very anxious to publish his book. The many difficulties and obstacles must have depressed him deeply."[40] Monk has called 1919 "perhaps the most desperately unhappy year of [Wittgenstein's] life".[41] Frege's rejection of the *Tractatus*, root and branch, played a significant role in this. Two days after he received Frege's first letter reacting to his manuscript, Wittgenstein wrote to Hermine that Frege's reply "depressed" him.[42] As Monk has put it, "there are some indications that it was Frege's response to the book that Wittgenstein most eagerly awaited. If so, the disappointment must have been all the more great when he received Frege's reactions".[43]

In the long, tense period of several months Wittgenstein was in captivity waiting to hear from Frege, the tension must have been nearly unbearable. He had written to Russell (on 12 June 1919), having not yet heard back either from the second publisher to whom he had turned (Braumüller, with the aid of a prior letter he solicited from Russell) or from Frege (he was to hear from Frege shortly, on 28 June; Braumüller and Frege's contacts were to reject the idea of publishing the manuscript.) Wittgenstein was at last sending Russell (with Keynes's assistance) a copy of his manuscript, partly exercised by anxiety about its ultimate worth and fate, and partly in response to remarks Russell had made in his *Introduction to Mathematical Philosophy* (sent to him by Russell earlier in the spring). There Russell set forth in print an account of what he said were some of Wittgenstein's views about logic.

[38] His frequently suicidal state in the later summer and early fall of 1919 are described by Monk, *Ludwig Wittgenstein: The Duty of Genius*, pp. 170ff.

[39] Again, compare von Wright's "The Origin of the *Tractatus*", especially pp. 77ff., and Monk, *Ludwig Wittgenstein: The Duty of Genius*, p. 170ff. for a discussion of Wittgenstein's initially fruitless efforts to have his essay published without subvention, which more than one person raised as a possibility (and he roundly rejected), and compare the discussion by McGuinness in *Wittgenstein: A Life, Young Ludwig 1889–1921*, pp. 267ff.

[40] Georg Henrik von Wright, "The Origin of the *Tractatus*", p. 78.

[41] Monk, *Ludwig Wittgenstein: The Duty of Genius*, p. 181.

[42] Wittgenstein to Hermine Wittgenstein, 1 August 1919, *Briefwechsel*, makes clear that he received Frege's letter on 30 July 1919. On 3 August 1919 Wittgenstein had written back to Frege, a letter that Frege did not reply to explicitly, on grounds that "it set so much in motion in me that if I had followed up on every stimulating point I would have had to write a book rather than a letter" (Frege to Wittgenstein 16 September 1919).

[43] Monk, *Ludwig Wittgenstein: The Duty of Genius*, p. 163.

Responding to this, Wittgenstein wrote to Russell in desperate frustration and anxiety that

> I'd very much like to write some things to you.—I should never have believed that the stuff I dictated to Moore in Norway six years ago would have passed over you so completely without trace. In short, I'm now afraid that it might be very difficult for me to reach any understanding with you. And the small remaining hope that my manuscript might mean something to you has completely vanished ... [The manuscript] is my life's work! *Now* more than ever I'm burning to see it in print. It's galling to have to lug the completed work round in captivity and to see how nonsense has a clear field outside! And it's equally galling to think that no one will understand it even if it does get printed!

And after receiving Frege's comments, on 6 October 1919 he wrote to Russell that

> I often feel miserable!—I'm in correspondence with Frege. He doesn't understand a single word of my work and I'm thoroughly exhausted from giving what are explanations pure and simple.[44]

As we know from correspondence surrounding later efforts to publish the *Tractatus*, Wittgenstein considered the whole idea of subventing the publication of his own work, directly or indirectly, through anything but its perceived philosophical merits, utterly humiliating and inappropriate.[45] Certainly by the end of the First World War Wittgenstein's whole attitude toward the making of such gifts, and the handling of money in general, had changed markedly: committing what was described as "financial suicide", he insisted on giving up any access to his family's fortune.[46]

Was this attitude toward his family's fortune *merely* "sentimental" or monkish? Wittgenstein's sister Hermine suggests in her recollections of his life that his change in attitude reflected a religious conversion which took place during the war, but even if such an awakening of religious feeling did color Wittgenstein decision, other explanations may be offered.[47] On the matter of

[44] See *Briefwechsel* and *Cambridge Letters*, pp. 131–132.

[45] Compare Wittgenstein's outraged comments about the publisher Braumüller's suggestion that he pay for the publication of the manuscript in a letter to Ficker of c. 7 October 1919; these and the relevant surrounding correspondence with Russell, Engelmann and others about such "humiliating conditions" are translated and discussed in von Wright, "The Origin of the *Tractatus*".

[46] See McGuinness, *Wittgenstein: A Life*, p. 278 and Monk, *Ludwig Wittgenstein: The Duty of Genius*, p. 171.

[47] See Hermine's contribution in Rhees, ed., *Recollections of Wittgenstein*, pp. 3–4. Her remarks should be compared with McGuinness's and Monk's biographical discussions, respectively, and with some remarks on asceticism in McGuinness's "Asceticism and Ornament", in his *Approaches to Wittgenstein: Collected Papers*.

money generally, we may see at work in Wittgenstein's decision a characteristic mixture of intellectual and practical motives. Given what we know of the vexations he faced in attempting to place his first manuscript, his decision to distance himself from the financial side of his life may also be viewed as a hardheaded and practical action, based as much on self-knowledge and an effort to quell and master anxiety as it was on an embrace of personal austerity, purity, and simplicity for their own sake. After all, had Wittgenstein retained any connection with the fortune and the family's decision-making regarding subvention of artists and intellectuals, he would have faced a constant stream, not only of distracting, anxiety-provoking and time-consuming administrative questions about the trust, but also public and private requests to help financially with bequests to particular intellectuals and institutions.[48] His family was, at his insistence, to protect him from this. Had they not done so, there would always have been questions, in his own mind and in others', about the extent to which his academic and intellectual recognition were a function of his family's position and notoriety. As Brian McGuinness has suggested (in conversation), had he stayed in Vienna, Wittgenstein faced the nearly certain fate of being constantly perceived and dimissed as nothing more than a wealthy amateur—and then facing his own reactions to this. In the end he escaped all this, severing to the greatest extent possible his connections to the family fortune and emigrating.[49] This did, at the very least, allow him more fully to concentrate on philosophy—even if it stoked the flames of a certain unhealthy vanity and self-isolation. Given his highly anxious nature, his nearly obsessive need to try to control how his thoughts were interpreted and received, and his equally obsessive counterbalancing struggle to let go entirely from concern with the fate and effects and perception of his writings and work, this decision may well have been a necessary condition for his philosophical productivity.

[48] Here it is useful to compare the correspondence between Wittgenstein's sister Gretl and Ludwig regarding Waismann's request, after Schlick's assassination in 1936, that the Wittgenstein family endow a professorship in Vienna in Schlick's name. Mining's report to Ludwig (in a letter of July 11, 1936) is that she was made very uncomfortable about this request, and told Waismann that "we" (i.e., the Wittgenstein family) "would never do such a thing", that "we are not influential, and, even if we were, we would never apply ourselves to such a thing, and even if we did, you would kill us, and even if you didn't, you would never allow such a thing to be considered" (see *Briefwechsel*).

[49] He did not fully succeed, given subsequent events following the *Anschluss* of Austria in March 1938, for large-scale decisions about the handling of the family fortune required a unanimous vote of the siblings. Monk details Ludwig's entanglement in the harrowing family battle over whether to hand over foreign currency to the Nazis in exchange for Aryan papers in *Ludwig Wittgenstein: The Duty of Genius*, p. 400. Compare Ursula Prokops's biography *Margaret Stonborough-Wittgenstein: Bauherrin Intellektuelle Mäzenin* (Vienna: Böhlau Verlag, 2003) as well as related correspondence in *Wittgenstein Familienbriefe*, eds. B. McGuinness, M.C. Ascher, and O. Pfersmann (Vienna: Hölder-Pichler-Tempsky, 1996) and in *Briefwechsel*.

Certainly in 1936 Wittgenstein would not have wanted the "personal" matter of his subvention of Frege—especially since it had occurred so near to the publication of his first manuscript—to be, as he wrote to Scholz, "set up in a public collection of Frege's work", especially given the criticisms of the *Tractatus* Frege documented in them. To move to the third reason he gave Scholz for not handing over the correspondence, this would have "perturbed" him. The public appearance of Frege's negative reactions to the *Tractatus* would have been likely to cost him time, trouble and emotional turmoil. His vanity and pride would have risked being set in motion knowing these letters to have been placed before the eyes of the public, thereby initiating a struggle, whatever actually transpired, with his own fears and anxieties about how his ideas and person were going to be received. He would have had to expect (or at least feared fearing to expect) that he would be asked to explain publicly why he had taken no account of Frege's criticisms in the *Tractatus* itself, but instead ignored them and pressed forward with the book's publication. This was especially sensitive for Wittgenstein in the 1930s, after he had changed his own thinking and yet continued to be perceived as an influential philosopher within the academy, constantly discussed and pressured for responses.

In early April 1936, at the time of the letter to Scholz and just before the assassination of Schlick (on 22 June), Wittgenstein was finishing the final year of his Trinity Fellowship. As Monk has described him, he had at this time

> ... little idea of what he would do after it had expired. Perhaps he would go to Russia—perhaps, like Rowland Hutt, get a job among 'ordinary people'; or perhaps, as Skinner had wanted, he would concentrate on preparing the *Brown Book* for publication. One thing seemed sure: he would not continue to lecture at Cambridge.
> ... [Wittgenstein had] doubts about his status as a philosopher, ... weariness of 'seeing queer problems' and [a] desire to start playing the game rather than scrutinizing its rules. His thoughts turned again to the idea of training as a doctor ... He suggested to Drury that the two of them might practice together as psychiatrists. Wittgenstein felt that he might have a special talent for this branch of medicine, and was particularly interested in Freudian psychoanalysis.[50]

At the time he wrote to Scholz, then, Wittgenstein was casting about in different directions for new paths in his life and thought, while at the same time still working up his philosophical ideas with an eye toward possible publication.[51] About the public appearance of any commentary or analysis of his ideas, Wittgenstein was, to put it mildly, extremely sensitive and liable to try to exert control where he could, especially when he was hard at work articulating and developing his views. The *Tractatus* had gained him his initial influence

---

[50] Monk, *Ludwig Wittgenstein: The Duty of Genius*, pp. 354 and 356.

[51] Compare Rhees's testimony, recounted in Monk, *Ludwig Wittgenstein: The Duty of Genius*, p. 357.

and reputation. But, at the same time, he had come to see it as flawed, both in its presentation of a conception of logic and also philosophically. He did not approve of its effects on the Vienna Circle, as he had repeatedly said to Schlick and to Waismann and indicated publicly in lectures such as the 1929 "Lecture on Ethics". As he had written in his diary in 1930,

> My book the *Log. Phil.Abhandlung* contains alongside good and genuine also Kitsch, that is, passages with which I filled up holes and so to speak in my own style. How much of the book consists of such passages I do not know and that is fairly difficult to assess.[52]

Feeling the continuing pressure and buzz about his reputation and ideas,[53] with an increasingly solid sense of how better to articulate his new philosophical ideas than in 1930, but aware that they were not yet formulated sufficiently well to be brought before the world in a book, Wittgenstein was at least honest with Scholz about his own emotional and intellectual state: he was neither intellectually nor emotionally prepared at this time to surrender these mementos to the eyes of the world.

From Wittgenstein's perspective, by retaining the letters with his papers, rather than Frege's, he would not be depriving the public of any useful ideas about his early works, though he would most certainly delay or perhaps ultimately suppress their publication. At the same time, this suppression would accomplish the not wholly unworthy aim of protecting him from being "perturbed" by public scrutiny and challenge, either of his work, his conduct in relation to Frege, or the publication of the *Tractatus*, a work he himself now considered to be flawed. By not destroying the correspondence, he would hold his cards and keep his options open, retaining it among his wider collection of correspondence. The cards and letters might or might not see the light of day later on, but Scholz's idea—publication and archiving of the letters in the context of Frege's scientific works—was not, in any case, the proper venue for them. How could Scholz, a theologian, philosopher and mathematical logician of a quite different stripe from himself, have been expected to understand what Wittgenstein had been attempting at the time of writing the *Tractatus*?[54] Frege, after all, had not! As Wittgenstein made clear to

---

[52] My translation; cf. entry of 16 May 1930 in *Denkbewegungen, Tagebücher 1930–1932/1936–1937 (MS 183)*, ed. I. Somavilla (Innsbruck: Haymon Verlag, 1997), p. 28 and in English, *Ludwig Wittgenstein: Public and Private Occasions*, J.C. Klagge and A. Nordmann, eds. (New York: Rowman & Littlefield Publishers, 2003), p. 39.

[53] On his philosophy of mathematics, this point is explained well in Ray Monk, "Bourgeois, Bolshevist or Anarchist? The Reception of Wittgenstein's Philosophy of Mathematics", in *Wittgenstein and His Interpreters*, eds. G. Kahane, E. Kanterian, and O. Kuusela (Malden, MA: Blackwell, 2007), pp. 269–294.

[54] In a letter to Oskar Becker of August 13, 1954, Scholz writes that the pages of a sketch he had worked up about Wittgenstein's later writings "that went out in the same mail which I sent to you, have been returned. I will not be agonizing

Scholz, he did not believe that in retaining the letters he was suppressing any material that would be appropriate for an edition of scientific correspondence of Frege's writings. Moreover, on his judgment, whatever intellectual value the letters contained formed a proper part of his own intellectual and "sentimental" development, rather than Frege's development as a logician. Their true significance lay not in their scientific or philosophical worth, but in their relation to the then unfolding history of the *Tractatus* and his own earliest efforts to explain and publish that work.

II

We have looked at Wittgenstein's 1936 answers to our main questions. But this is not the only relevant point of view. Returning to more detailed consideration of Frege's criticisms of the *Tractatus* will help to better weigh their possible philosophical significance. In particular, as Janik suggested, we need to ponder the role of the concept of clarity as it figures, both in Frege's reactions to Wittgenstein and in Wittgenstein's to Frege, as well as in our understanding of how certain lines of thought emerged within early analytic philosophy.

Like Janik, I take the letters to provide us with an emblem or lesson about the difficulty of reaching agreement about what philosophical "clarity" in one's thought and expression requires, even and perhaps especially between thinkers who take themselves to be devoted in special measure to achieving it. The letters do confirm, it seems to me, that one of the most central and lasting formative impulses in early analytic philosophy was a preoccupation, not with positivism and verificationism about meaning and necessity, but rather with the complexity and unclarity of the notion of analysis itself, that is, with challenges facing philosophical accounts of what it is for thought and truth to be clearly expressed in language, and what the role, status, and contributions of logic and of symbolism are in meeting them. Frege and Wittgenstein do have different, perhaps even "utterly distinct" conceptions of how we are to view the outcome and goal of logical clarification, but we must remember that their devotion to the purposes and value of this kind of clarification, and their sense of the range of possible answers to questions about the basic notions of logic, is shared. Within their departures from one another lies then a large region of overlap, as Frege's letters seem to attest: Frege repeatedly emphasizes his hope of reaching agreement with Wittgenstein in those areas—well realizing, after he saw the manuscript of the *Tractatus*, that there would remain a philosophical penumbra where there could be no meeting of minds.

---

any more about it. These pseudo-sibylline pages have absolutely nothing in them for me". (The letter is in the Scholz archive at the University of Münster library, along with correspondence with von Wright in which Scholz is open about his inability to make headway with Wittgenstein's writings, or with any philosophy inspired by it.) Wittgenstein's 1936 brush-off may or may not have led to Scholz's later frustration.

A primary theme in the *Tractatus* is an investigation of what is involved in the idea of representation of reality—an investigation whose coherence Frege explicitly rejects as fundamental to logic, both in his letters to Wittgenstein and, more explicitly than in any other essay he wrote, in "Der Gedanke". Now quite apart from Frege's reading of the *Tractatus*, it must be said that gauging the ultimate aims of Wittgenstein's *Tractatus* discussion of pictures and representation is a challenge. I believe that Wittgenstein is partly involved in a critical rethinking of the Idealist, i.e., Kantian tradition, in which the notions of *form*, *idea* and *representation* figured centrally. His conception of sentences as pictures also fashions a critical response and alternative to Moore's and Russell's (various) accounts of truth and propositionhood, each of which bypassed the notion of representation altogether.[55] More directly at issue in relation to the correspondence with Frege is the question of how far the *Tractatus* does and does not offer views consistent with, or at least coincident to, Frege's.

For some readers of the *Tractatus* (not myself, but perhaps for Frege, and certainly for some later readers of the book) Wittgenstein's conception of propositions as models of reality should be taken as a "theory", perhaps even a correspondence theory, of truth (or perhaps of meaning). Frege explicitly argues in the opening pages of "Der Gedanke" that any such theory is incoherent. Hans Sluga has gone so far as to claim that Frege wrote "Der Gedanke", in particular its criticisms of correspondence theory, "with Wittgenstein in mind", stimulated by the manuscript of the *Tractatus* "to give his views a final and definitive airing before Wittgenstein could lay out his related though distinct ideas".[56] And it is true that at the outset of the essay Frege criticizes the idea that truth is a property of representations or pictures or facts. Let us here set aside the interesting yet murky question of how the *Tractatus* might have influenced Frege's latest writings. This is possible, perhaps even likely, though difficult to pin down. I note that it is a corollary of Sluga's view that Frege failed to appreciate what Sluga also calls Wittgenstein's proceeding, after the early parts of the *Tractatus*, "to deconstruct all semantic theorizing" and to "conclude that all attempts to speak *about* logic are bound to fail".[57] This outcome, for Sluga, makes Wittgenstein's *Tractatus* views similar to Frege's own later views on the primacy of judgment, or recognition-of-truth, for logic. But neither Frege nor Wittgenstein saw things this way.

An alternative or perhaps supplementary interpretation would emphasize that Wittgenstein's conception of sentences as pictures serves, not only as a theory or a preliminary step in deconstructing the correspondence theory of truth, but instead to tame and incorporate into Wittgenstein's way of think-

[55] This is discussed in Thomas Ricketts, "Pictures, Logic, and the Limits of Sense in Wittgenstein's *Tractatus*", in H. Sluga and D. Stern eds., *The Cambridge Companion to Wittgenstein* (New York: Cambridge University Press, 1996), pp. 59–99.

[56] Hans Sluga, "Frege on the Indefinability of Truth", in E. Reck, ed., *From Frege to Wittgenstein* (op. cit.), pp. 75–95; quotations from pp. 89, 77.

[57] Sluga, "Frege on the Indefinability of Truth", p. 92.

ing legitimate elucidatory talk of correspondence, facts and situations. Here the remarks treating sentences as pictures are intended to emphasize that sentences themselves are facts, understood as perceivable symbolic structures placed within a "space" of form, i.e., a system of representation that we use. This brings out an holistic strand in Wittgenstein's philosophy of sentences that resonates directly with the Fregean context principle, a principle quoted in the *Tractatus* (at 3.3) and clearly of importance for Wittgenstein's philosophy of logic.[58] Of course, unlike Frege Wittgenstein refuses to see sentences as proper names of truth-values (their *Bedeutungen*) which simultaneously express a separate level of sense, or thought: his picturing conception is intended to avoid the dualism of levels of meaning, letting the sentence, like a picture, express its sense off its own bat, so to speak. On this reading a substantive correspondence theory of truth is never at issue between Wittgenstein and Frege, despite the concerns Frege expressed at the language of "facts" in his letters.

Whether the *Tractatus* conception of sentences as pictures is viewed as a theory or not, it is clear that it helps to set up Wittgenstein's own treatment of logical form as non-picturing, and logic as non-factual. Thus Wittgenstein's conception of sentences as models of reality does not undercut, but reinforces his central concern, not only with the importance and nature of symbolism to logic, but also with the need for the sorting out and distinguishing different dimensions or roles of expression in connection with our uses of symbolism in logic. This is indeed a Fregean, as well as a Russellian theme. But in the *Tractatus* the sorting out is framed by Wittgenstein's distinctive preoccupation with a question that neither Frege nor Russell had pursued, viz., "What is the nature of the logical as such?"

For Frege the notion of recognition-of-truth in judgment is basic to a proper understanding of logic,[59] whereas for Wittgenstein logic's sole concern is with clarifying, through rearrangement of our expression, what it is for sentences to express senses, true or false. One of the chief philosophical aims of the *Tractatus* is to show how a marking off of that which distinctively belongs to the essence of logic requires clarification of the very idea of propositions as representations of reality, true or false—and vice versa, since on his view logic is an activity *rather than* a body of propositions, true or false.[60] There are no

[58] On the change between Wittgenstein's earliest writings up through the *Prototractatus* to the more "holistic" use of Frege's context principle in the *Tractatus*, see Michael Kremer, "Contextualism and Holism in the Early Wittgenstein", *Philosophical Topics* 25, 2 (1992): 87–120.

[59] See Thomas Ricketts, "Logic and Truth in Frege", *The Aristotelian Society Supplementary*, 70 (1996): 121–140.

[60] On the importance of faithful representation of reality to ideas in the *Tractatus*, see Hintikka, "What Does the Wittgensteinian Inexpressible Express?", *The Harvard Review of Philosophy*. I reply to some of Hintikka's views in my "Wittgenstein and the Inexpressible" in A. Crary, ed., *Wittgenstein and the Moral Life: Essays in Honor of Cora Diamond* (Cambridge, MA: MIT Press), pp. 177–234.

logical facts, according to the *Tractatus*, and no logical propositions: logic is not a science of any kind aimed directly at truth or facts, for the logic of the facts cannot itself be represented in the same sense as facts are. Wittgenstein thus required a distinctive conception of the factual and of representation to work out his distinctive conception of logic. In the *Tractatus*, self-reflexively but fully consistently, he treated these notions and distinctions as themselves logical, what he called "formal", conceiving of his remarks as elucidatory or exhortatory, rather than strictly speaking scientific. His conception of "form" or symbolic structure as elucidated through possibilities of rearrangement of expression allowed Wittgenstein retain his ties to the logicist idea of logic as universally applicable, constitutive of our understanding of content. For "formality" did not mean for him, as it had for Boole and the algebraists of logic, an emptiness of content understood as an open-ended conception of the reinterpretability of empty signs.

Frege's mature philosophy of logic—as expressed, for example, in "Der Gedanke"—also serves to attempt to liberate logic from the notion of *fact*, but differently, for Frege always viewed logic as a science. Frege's conception of logic rests on a primitive notion of recognition-of-truth, and in "Der Gedanke" he uses this conception to argue explicitly against the definability of truth, the correspondence theory of truth (whether framed in terms of facts or not), and more generally the idea of truth as a genuine property (e.g., of pictures or of sentences). By contrast, it is clear that in the *Tractatus* framework recognition-of-truth could play no role in logic at all—as opposed to the notion of sentences as symbols expressing senses or thoughts, i.e., sentences, true or false. Frege wishes to resist the reduction of thoughts to sentences; this is why he speaks of thoughts as inhabiting a "third realm" in "Der Gedanke", a realm whose structure we acknowledge and express in recognition-of-truth. (Already in his letter following his 9 November 1913 meeting with Wittgenstein Frege had complained that "W. places too great value upon signs".[61]) By contrast Wittgenstein, who had thought through Russell's emphasis on a theory of symbolism, takes the notions of sense and thought to belong to sentences as symbols, i.e., signs whose uses contribute to the expression of propositions, true or false. Wittgenstein's treatment of sense as expressed in the "bi-polarity" of sentences (their being true or false, depending upon how the facts are) is designed to reject Frege's two-tiered view of sense and reference, both for propositions and for proper names. It brings into view a notion of facts standing outside their particular form of representation (an anti-Idealist element) and presents a view of logic on which there are no logical laws.

What is thus most philosophically significant about the letters is that Frege focuses in the majority of his substantive remarks on the notions of "fact" and "atomic fact", especially on the idea that these *correspond* to a true sentence,

[61] See my précis of the Scholz list comments in my Preface to the translations of the Frege-Wittgenstein correspondence in this volume.

or exist if a sentence of the appropriate form is true. Here he is concerned to question whether this language can contribute to useful elucidation of fundamental logical notions. His resistance to treating phrases such as "is a fact" and "what is the case" as informative explications of truth or as basic to our understanding of logic had been longstanding, but by the time he read the *Tractatus* manuscript, as he was finishing "Der Gedanke", the resistance was in full flower. Thus he repeatedly emphasizes to Wittgenstein the logical structure and role of definitions as replacing whole, complex expressions, finding the language of the opening remarks of the *Tractatus* connecting the notions of *fact*, *state of affairs*, and *atomic fact* lacking in "sufficiently detailed" explicit justification and elucidation of primitive notions through logical segmentation. (One is reminded, in reading Frege's questions to Wittgenstein about the notion of a "constituent" of a fact, of his earlier correspondence with Russell; it is tempting to surmise that he read Wittgenstein's remarks as simply rewarming old Russellian ideas, rather than reconceiving their role and significance.[62]) His remarks should therefore also be understood against the backdrop of his own development and the arguments he made against certain conceptions of "existence" and "truth" in his later writings, of which "Der Gedanke" is one.[63]

I remarked above that whereas the correspondence shows the *Tractatus* to have brought about essentially no evolution in Frege's views, philosophical parts of the correspondence do appear to have remained with Wittgenstein long after 1920. I turn next to this theme.

The final letter of the correspondence (3 April 1920) squares with an anecdote of Geach's recounted 11 years before its discovery in 1989. In "the final months" of Wittgenstein's life, Geach had written,

---

[62] See Russell's letter to Frege of 12 December 1904 in Frege, *Philosophical and Mathematical Correspondence*, eds. B. McGuinness, G. Gabriel et al., trans. H. Kaal (Blackwell/University of Chicago Press, 1980), especially p. 169. Goldfarb ("Wittgenstein's Understanding of Frege", p. 188) says he knows of no evidence that Wittgenstein discussed Frege's work with Russell (nor do I). But it is difficult to imagine that the subject of Frege on sense and reference never came up.

[63] Readers may see Sluga, "Frege on the Indefinability of Truth" for an analysis of Frege's own evolution with regard to the notion of truth. With respect to Wittgenstein's development, Goldfarb argues persuasively that at least in the pre-*Tractatus* writings "the priority for Frege of the notion of recognition-of-truth to that of truth did not register on Wittgenstein, or at least there is no evidence that it did ... Frege elaborates the point only in "Thoughts" ... and in unpublished writings" ("Wittgenstein's Understanding of Frege", p. 192). What I am arguing here is that given Wittgenstein's views on the nature of logic, which were after all in place well before the manuscript of the *Tractatus* was written, it would not have been possible for him to agree with Frege's idea of recognition-of-truth as a basic logical notion. I fully agree with Goldfarb that Frege's conception cut off at the pass, as perhaps Wittgenstein's did not, the very idea of facts or configurations that render our propositions true, and the Frege-Wittgenstein correspondence seems to confirm this.

> [Wittgenstein] took a good deal of interest in the plan Max Black and I had for a little book of Frege translations; and it was through him that I was able to locate some rare works of Frege—the review of Husserl's *Philosophie der Arithmetik* and the essays 'Was ist eine Function?' and 'Die Verneinung'—in the Cambridge University Library. He advised me to translate 'Die Verneinung', but not 'Der Gedanke': that, he considered, was an inferior work—it attacked idealism on its weak side, whereas a worthwhile criticism of idealism would attack it just where it was strongest. Wittgenstein told me he had made this point to Frege in correspondence: Frege could not understand—for him, idealism was the enemy he had long fought, and of course you attack your enemy on his weak side.[64]

Wittgenstein's sharing his recollection with Geach had its effect: Geach and Black did not include a translation of "Der Gedanke" in their influential collection *Translations from the Philosophical Writings of Gottlob Frege*, whose first edition appeared in 1952. (Geach's translation of "Der Gedanke" and his publication of Wittgenstein's testimony awaited the publication of Geach's much later 1977 edition of Frege's *Logical Investigations*, the Preface of which contains the above-quoted passage).[65] And it is surely relevant to the question of Wittgenstein's later attitude toward Frege's criticisms of the *Tractatus* that Wittgenstein insisted to Geach, just as he had to Frege thirty-odd years before, that "Der Gedanke" was an inferior work because it missed the *logic* of idealism—attacking it, Wittgenstein said, "on its weak side", thereby missing the "deeper grounds" of idealism, its "deep and true core", "an important feeling that is wrongly gratified, hence, a legitimate need" (cf. Frege to Wittgenstein 3 April 1920). Frege had asked in reply, "Of what sort is this need?", insisting that apparent grounds are not grounds at all, and that it was no part of his intention "to trace all ... disturbances of psychologico-linguistic origin" leading to philosophical error (cf. Frege to Wittgenstein 3 April 1920).

Geach suggests that

> ... in spite of Wittgenstein's unfavourable view of 'Der Gedanke', his later thought may have been influenced by it. It would not be the only time that Frege's criticism had a delayed action in modifying Wittgenstein's views after he had initially rejected the criticism.[66]

And Geach forwards two examples of this "delayed action". Let us consider them in turn.[67] First,

---

[64] Geach, Preface to Frege, *Logical Investigations*, ed. P.T. Geach, trans. P.T. Geach and R.H. Stoothoff (New Haven: Yale University Press, 1977), p. vii.

[65] This point is made by Erich Reck, in his "Wittgenstein's 'Great Debt' to Frege", p. 27.

[66] Geach, Preface to Frege, *Logical Investigations*.

[67] Both examples are from Geach, Preface to Frege, *Logical Investigations*.

> Wittgenstein told me how he had reacted to Frege's criticism of the Russellian doctrine of facts—a doctrine still presupposed in the *Tractatus*. By this view, such a fact or complex as knife-to-left-of-book would have the knife and the book as parts—though Russell of course avoided the rude four-letter word 'part' and spoke of constituents. Frege asked Wittgenstein if a fact was *bigger* than what it was a fact about; Wittgenstein told me this eventually led him to regard the Russellian view as radically confused, though at the time he thought the criticism silly.

It is difficult to know how to weigh this suggestion insofar as it has a bearing on the *Tractatus* and Frege's correspondence about it with Wittgenstein; unlike the subsequent example we shall consider, Geach does not report Wittgenstein saying explicitly that the correspondence dealt with it. The difficulty is that it is unclear, at least for many readers, how and in what way (if any) the *Tractatus* is committed, as Russell once was, to a "doctrine of facts" that takes constituents of facts to be objects existing prior to any particular analysis of the language. Moreover, it is unclear when Frege made this objection to Wittgenstein, and when we are to suppose Wittgenstein became moved to think it something better than "silly". The objection as described does not explicitly occur in Frege's letters reacting to the manuscript of the *Tractatus*, and it is difficult to see how it could have had such a profound "delayed" reaction if Wittgenstein is supposed to have (1) thought so highly of it as a cogent critique of his book and yet (2) never once in his manuscripts (which often mention Frege) mentioned it. The objection would have had to be made *prior* to Wittgenstein's writing of the *Tractatus*, and Wittgenstein would have to be supposed to have ignored it altogether, but later on come to appreciate its force. But we have no record of this apart from Geach's anecdote.

The closest relevant remark in the correspondence is one in which Frege trots out a line of thought he must have associated with a *Russellian* view of constituents. For his example of Vesuvius reminds us of Russell's example of Mont Blanc, which Russell offered to Frege as part of an objection to the Sinn/Bedeutung distinction in a letter of 12 December 1904. Russell had written to Frege that

> Concerning sense and meaning, I see nothing but difficulties which I cannot overcome. I explained the reasons why I cannot accept your view as a whole in the appendix to my book [*The Principles of Mathematics*], and I still agree with what I there wrote. I believe that in spite of all its snowfields Mont Blanc itself is a component part of what is actually asserted in the proposition 'Mont Blanc is more than 4000 metres high'.[68]

---

[68] See Russell to Frege of 12 December 1904, in Frege, *Philosophical and Mathematical Correspondence*, p. 169.

Puzzled by Wittgenstein's adherence to the language of "facts", "states of affairs" and "atomic facts" in the *Tractatus* manuscript, in Frege's initial reply (the 28 June 1919 letter) he presses Wittgenstein on the question of what "binds" the constituents of a fact together, asking "Can this perhaps be gravitation, as with the system of planets?" This is a pointed question surely intended to be understood as a *reductio* of the whole way of thinking. Frege pursues the point, saying that without *examples* of atomic facts, facts, things and states of affairs, or some clarification of what corresponds linguistically to these notions,

> it appears that constituents of Vesuvius must also be constituents of this [atomic] fact [about Vesuvius]; the fact will therefore also consist of hardened lava. That does not seem right to me.

This conjures up the spectre of a view like Russell's, in which the parts of the mountain itself are parts of that which is (asserted) in a proposition. Frege is here asking Wittgenstein to clarify the status of his *Tractarian* distinctions. And it is possible that this is the criticism which Geach reports Wittgenstein having said had a "delayed reaction" on his thinking. For *if* it makes sense to say that the fact about Vesuvius is itself made partly of lava (Frege's question to Wittgenstein, inspired, he writes, by *Tractatus* 2.011), *then* it would make sense to ask whether "a fact was *bigger* than what it was a fact about"—whether, so to speak, whatever is predicated of Vesuvius is included in the fact as well, as a constituent or thing. Yet if Wittgenstein's whole point in the *Tractatus* is to show the "formality" of the interrelated notions of fact and situation (the hopelessness of framing propositions about them, true or false), then Frege is missing his point. It is certainly true that Wittgenstein later on became highly disillusioned with the *Tractatus*'s willingness to truck in the Russellian language of facts, states of affairs and their constituents, and so on. In particular, he complained that he had failed to give *examples* of simple objects in his book, while insisting at the same that that *there must be such*.[69] Possibly, Frege's correspondence, in which the absence of examples is explicitly complained of, played a role here. Nevertheless, it is difficult to imagine that the question Geach reports Frege having posed to Wittgenstein had on its own a singularly powerful "delayed action" on Wittgenstein, even if we grant the full accuracy of Geach's and Wittgenstein's recollections: there were too many other difficulties for Wittgenstein (and for Frege) to have had with the book. Indeed, it seems just as likely, so far as I can see, that the objection Wittgenstein recalled Frege making was offered to him much earlier, in their discussions before the war, and had already had its effect even before Wittgenstein wrote the *Tractatus*.

[69] This is reported by Norman Malcolm, in his *Ludwig Wittgenstein: A Memoir* (2nd edition, New York: Oxford University Press, 1984), p. 70.

Geach offers a second example of the "delayed action" of a criticism of Frege's, this one contained in "Der Gedanke" and tied explicitly to the correspondence:

> ... In 'Der Gedanke' Frege lays down premises from which it is an immediate consequence that certain ideas he plays with in the essay—private sensations with incommunicable qualities, a Cartesian *I* given in an incommunicable way—are really bogus ideas, words with no corresponding thoughts. For Frege affirms (1) that any thought is by its nature communicable, (2) that thoughts about private sensations and sense-qualities and about the Cartesian *I* are by their nature incommunicable. It is an immediate consequence that there can be no such thoughts. Frege never drew this conclusion, of course—even though the passage about the two doctors, for whom the patient's pain can be a common object of communicable thoughts without their needing to *have* the pain, comes close to the rejection of pain as a private incommunicable somewhat. But though he never drew this conclusion, Wittgenstein was to draw it.[70]

Just how Wittgenstein supposed a truly *proper* critique of idealism was to proceed, as opposed to an attack on it "where it is weakest"—which is what he took "Der Gedanke" wrongly to offer—is a fascinating and, I believe, as yet still unresolved interpretive question about the *Tractatus*, not merely about Wittgenstein's later philosophy.[71] It is interesting that in his final letter to Wittgenstein (3 April 1920) Frege raises the issue twice, alluding to Wittgenstein's earlier remarks. Here I believe we learn something, not merely about the later, but also the early Wittgenstein. For we may infer at least this much from the exchanges and reports: not only in later life, but even in the *Tractatus*, Wittgenstein attempted, not merely to reject, but to represent and do justice to idealism, to show how and why the *logic* of idealism (or, equivalently here, scepticism) has a "deep and true core", rooted in "an important feeling that is wrongly gratified". As he was later to emphasize, one of the most important tasks in philosophy "is to express all false thought processes so characteristically that the reader says, 'Yes, that's exactly the way I meant it"'.[72] This, as Frege wrote in his reply, had been no part of Frege's task. But

[70] Geach, Preface to Frege, *Logical Investigations.*

[71] I have tried to engage the structure and text of *Tractatus* with systematic aspects of the Idealist tradition in my essays "Tautology: How Not to Use a Word" (with B. Dreben), *Synthese* 87/1 (April 1991): 23–50 and "The Uncaptive Eye: Solipsism in the *Tractatus*" in L. Rouner, ed., *Loneliness* (Notre Dame: Boston Studies in the Philosophy of Religion, 1998), pp. 79–108. See also David Pears, *The False Prison*, vol. I (New York: Oxford, Clarendon Press, 1987) and Peter Sullivan, "The truth in solipsism, and Wittgenstein's rejection of the *a priori*", *European Journal of Philosophy* 4 (1996): 195–219.

[72] See TS 213, pp. 405–435 of *The Big Typescript*, eds. and trans. C.G. Luckhardt and Maximilian A.E. Aue (Malden, MA: Blackwell Publishing, 2005).

apparently it was acknowledged by both of them to form part of Wittgenstein's in the *Tractatus.*

Recent suggestions of Ray Monk, who has written an introduction to Wittgenstein's thought, contribute in a different way toward our understanding of the correspondence's specific philosophical significance. Monk emphasizes the intrinsically enigmatic and difficult, perhaps insuperable, difficulties facing any interpreter of the *Tractatus.* This is useful to bear in mind if only because we need to remember that Frege, back in 1918–1919, writing before the main developments in the tradition, may be forgiven for having had trouble understanding it. Has *anyone* made sense of the book—except by rejecting large portions of its letter and spirit? This is doubtful. Clear it is not, as Frege repeatedly points out to Wittgenstein in the final four letters of the correspondence. As Monk aptly writes, of Wittgenstein's famed invocations of showing vs. saying (controversy about which has surrounded the book from the very beginning),

> The ongoing debate about the saying/showing distinction and about whether or not Wittgenstein thought it was possible to show philosophical truths through nonsensical propositions is just one among many controversies that divide interpreters of *Tractatus-Logico-Philosophicus.* And these controversies do not concern details but the very fundamentals of the book. More than eighty years after it was published, and despite a vast secondary literature inspired by it, there is still no general agreement about how the book should be read. It is surely one of the most enigmatic pieces of philosophy ever published: too mystical for logicians, too technical for mystics, too poetic for philosophers and too philosophical for poets, it is a work that makes extraordinarily few concessions to the reader and seems consciously designed to elude comprehension.[73]

Was the *Tractatus consciously* "designed to elude comprehension"? The closing lines of the *Tractatus*, in which Wittgenstein wrote of the nonsensical status of his remarks, have suggested that in some way this is so:

> 6.54 My propositions are elucidatory in this way: he who understands me finally recognizes them as nonsense [*unsinnig*], when he has climbed out through them, on them, over them. (He must so to speak throw away the ladder, after he has climbed up on it.)
> He must overcome [*überwinden*] these propositions; then he sees the world rightly.
>
> 7 Whereof one cannot speak, thereof one must be silent.

But the route to this overcoming, its method, presuppositions, means, and purposes, remain a source of fundamental controversy, as Monk says, among

[73] Monk, *How to Read Wittgenstein*, pp. 29–30.

readers of the *Tractatus*. This is not the place to survey recent twists in this controversy that have led to the debates between "new" and "old" readers of the *Tractatus* with regard to the topics of saying vs. showing, nonsense vs. sense, realism and idealism in the *Tractatus*. The Frege-Wittgenstein correspondence sheds little direct light on these issues, if only because Frege offered no sustained examination of Wittgenstein's *Tractarian* remarks about them.

McGuinness has suggested that Wittgenstein's aesthetic "asceticism"—his resistance to charm and ornament, whether in furniture, architecture, or literature, in his giving away of his money to live "simply", or in the unadorned structural organization of the *Tractatus*'s numbered remarks—reflects, at least in part, "the negative aim" of much of his philosophical work. This had implications, as McGuinness sees it, for the way Wittgenstein wrote and thought about himself:

> For him style, the way something was put, was of enormous importance, and that not only in the artistic sphere. He said once, it wouldn't matter what a friend had done but rather how he talked about it. Similarly he used to insist on a careful reading of the dictum, *Le style c'est l'homme meme*: the thought is that the real man reveals himself in his style. The meaning of the words, the content, is something secondary, and so likewise is the brute action performed. Of course, it is an important philosophical observation that actions cannot be separated from the way in which they are judged by him who performs them. Still there are dangers, if a feeling for style becomes the supreme commandment. It is not to be thought of that this was a risk for Wittgenstein in the moral sphere, but in aesthetics [as he himself suggested], he perhaps incurred it.[74]

Pursuing this thought in light of some of Wittgenstein's own self-criticisms, McGuinness points to what he takes to be "distortions" in Wittgenstein's later writings produced by his frequent (often alternative) draftings of his remarks, using multiple revisions of emphasis, underlining, and so on. For McGuinness, "the excessive frequency of accidentals in his manuscripts and typescripts"[75] reflects

> an almost pathological insistence on finding the correct distribution of emphasis in a sentence ... It is almost as if he regarded something as false as soon as it was written down ... It is not surprising, therefore, that Wittgenstein was profoundly dissatisfied with the accounts of his work that others gave ... Partly this is due to the negative aim of this work. It is intended to drive out the evil spirit from the reader as from his pupils. False philosophy must be exorcized. But that is an operation best performed *viva voce* and through personal contact.

[74] McGuinness, "Asceticism and Ornament", pp. 21–22.
[75] McGuinness, "Asceticism and Ornament", p. 22.

> One false notion is driven out, and immediately the next false notion that threatens to take its place must be corrected. A book or an article freezes what ought really to be a living flow of ideas.[76]

What resulted was a difficult question about which Wittgenstein himself at times worried: was he a "merely reproductive" thinker, *merely* an improver or trimmer of other's ideas, *merely* redistributing emphases in his sentences? McGuinness asks, partly on Wittgenstein's behalf, an excellent question:

> Was his philosophy bare asceticism without positive content to make it worth the effort and the abnegation? This difficult question must be resolved in any attempt to assign Wittgenstein a place in the history of ideas.[77]

In his recent remarks on Wittgenstein, Monk's resolution of this "difficult" question is clear: he extracts something more positive, even from the *Tractatus*, and precisely on the basis of Wittgenstein's aesthetic concerns and aspirations. As Monk sees it, Wittgenstein's concern with proper expression represented a devotion to authenticity, to presenting his ideas in a way that would not be "counterfeit", rather than an excessive tendency to pick at emphases.[78] Moreover, at the heart of the *Tractatus* lies, according to Monk's reading, an important "insight", one that would be differently articulated, though retained, in Wittgenstein's later writing, namely, that "philosophy ought to be written only as a poetic composition".[79] He points out that Wittgenstein's first choice of a publisher for his manuscript, the firm of Jahoda, was "not an academic publisher but a literary one, best known as the publisher of the Viennese satirist Karl Kraus";[80] Frege was resorted to only after that route had been blocked to him.

By "poetic composition" Monk has in mind what he takes to be a Wittgensteinian contrast between the value and aims of poetry, art, music and philosophy, and the value and aims of science. Wittgenstein had, after all, written in the *Tractatus* that the purpose of his book is to give an "understanding" reader "pleasure", that philosophy is not one of the natural sciences—that is, it is not a body of doctrine but an activity consisting essentially of "elucidations". Its results are then "not 'philosophical propositions' ", but instead "the clarification (*klar machen*) of propositions" (see *Tractatus* 4.11–4.112). One of Monk's interpretive ideas is that the *Tractatus*'s primary failure, as a work, was its failure, within the form of its "icy rigor of numbered propositions",[81] to make this distinction between philosophical and scientific understanding clear. The book's overarching structure and basic ideas about logical

[76] McGuinness, "Asceticism and Ornament", pp. 23ff.
[77] McGuinness, "Asceticism and Ornament", p. 24.
[78] See *Ludwig Wittgenstein: The Duty of Genius*, p. 346.
[79] Monk, *How to Read Wittgenstein*, p. 65.
[80] Monk, *How to Read Wittgenstein*, p. 30.
[81] Monk, *How to Read Wittgenstein*, p. 65.

form sinned against its best overarching strand, which was, at least for Monk, Wittgenstein's determination to resist scientism precisely by highlighting the integrity, autonomy and intrinsic value of non-scientific forms of understanding such as are found in philosophy and in poetry. Wittgenstein's insistence that the status of his own *Tractarian* remarks is that they are mere "elucidations", and "nonsensical" was, for Monk, "an obviously unsatisfactory evasion" of a central difficulty with his numbered style and method, a method that positively invited Waismann and others in the Vienna Circle (among others) to try to summarize its apparently theoretical doctrines about logic with a set of scientific-world-view "theses".[82]

While the "insight" into the value of non-scientific forms of understanding is not one I would want to deprive us of, and while I fully agree with Monk that the poetic qualities of Wittgenstein's writing are internal to its intellectual aspirations, it seems to me worth also emphasizing that at the time of writing the *Tractatus* Wittgenstein was possessed by a vision of a kind of unity between the activities of logic, philosophy and poetry—a vision that was, in one way or another, to continue to be reflected well into his later writings. Even if Wittgenstein's remarks about logical form in the *Tractatus* were later to dissatisfy him, it is important how it was that he conceived of the role of logic, for this conception stayed with him throughout his later life. Logic is depicted in the *Tractatus* as a way of coming properly to appreciate the importance of punctuation and/or syncopation in the presentation of thinking, an activity involving at its heart a progressive rearrangement of expressive elements of our language. Logical operations and even number words are explicitly held to be properly conceived of as expressed by punctuation marks, not constants (*Tractatus* 5.4611), not as elements of sentences having *Bedeutung* (Wittgenstein's *Grundgedanke* (5.4)). Logical axioms and laws are not necessary to logic's formulation: insofar as they clarify anything, they too are to be conceived as a style of presentation rather than an unearthing of fundamental representational truths or constants (6.127). Logic, philosophy, as Wittgenstein had come to stress early on, is not to be conceived of as part of natural science, but as a kind of activity of clarification, exposure of tautologousness and non-tautologousness, nonsense and sense. There is, as he had written earlier in his pre-war *Notebooks*, no need for a *theory* of symbolism, there is only symbolizing.[83]

From this perspective, the most interesting point to note about the Frege-Wittgenstein correspondence is that Frege immediately turns a hostile ear to the very idea of logic (or philosophy) as empty, as if instinctively grasping that this is where his confusions with the *Tractatus* (and his differences with

[82] See Monk, *Ludwig Wittgenstein: The Duty of Genius*, pp. 296–297. For remarks invoking a similar vision of what is most valuable in Wittgenstein's work, compare Putnam's remarks on the *Tractatus* in "Floyd, Wittgenstein and Loneliness", in L. Rouner, ed., *Loneliness*, pp. 109–114.

[83] I do not mean here that there was no development in Wittgenstein's views, as I make clear in my "Wittgenstein and the Inexpressible".

Wittgenstein) really lie. He is not deaf to Wittgenstein's poetic aspirations, but sees in them a profound difference between his self-conception and those of the author of the *Tractatus*

To see this, we should focus on the second of Frege's four letters on the *Tractatus*, that is, what he writes to Wittgenstein on 16 September 1919, in reply, not only to the *Tractatus* itself but to what must have been Wittgenstein's two rather desperate letters to him from Cassino (now sadly lost) attempting to clarify things. Frege's initial remarks on the manuscript (given in his letter of 28 June 1919, not received by Wittgenstein toward the end of July) had openly professed a lack of comprehension, and this had badly "depressed" Wittgenstein, as he had written to Hermine. We may surmise that in his initial two replies to Frege (the first of which was sent within 48 h of receiving Frege's letter (see footnote 42)) Wittgenstein tried to set Frege straight about what his poetic aims and purposes had been, as well as his existential state. He seems to have expressed doubt that they would ever be able to understand one another.

Frege says that he will "not so easily surrender the hope of reaching agreement with you", aiming to quell Wittgenstein's desperation. He thus holds out hope for an ultimate understanding, and, mentioning explicitly "the consequences of everything you had to go through" (during the war), attempts to reassure Wittgenstein about how well he thinks of him philosophically (this leads us to suppose that Wittgenstein had, as in his earlier letter to Russell, expressed doubt about this). Frege makes it clear that he hopes to learn from Wittgenstein and for Wittgenstein to learn from him, that he wishes to enter into a debate in which Wittgenstein will be "won over" to his point of view.

At this point he confesses that

> What you write me about the purpose of your book strikes me as strange. According to you, that purpose can only be achieved if others have already thought the thoughts expressed in it.

We may plausibly assume that Wittgenstein had called Frege's attention to remarks to this effect in the Preface of the *Tractatus*. Frege continues:

> The pleasure of reading your book can therefore no longer arise through the already known content, but, rather, only through the form, in which is revealed something of the individuality of the author. Thereby the book becomes an artistic rather than a scientific achievement; that which is said therein steps back behind how it is said. I had supposed in my remarks that you wanted to communicate a new content. And then the greatest clarity [*Deutlichkeit*] would indeed be the greatest beauty.

Frege's last sentence quotes a line of Lessing's from *Das Testament Johannis*, well-enough known that it is likely Frege would have assumed Wittgenstein to have heard it: "The greatest clarity [*Deutlichkeit*] was to me always the greatest beauty". In Frege's typically acute way, he lays down a gauntlet: either the manuscript is written in the spirit of a philosophical contribution toward clarification, hence, furthering a cognitive advance, or it is not. If it is, then Lessing's aesthetic remark placing the emphasis on communication and interpretability would apply. If not, then another aesthetic might be appropriate. But the logical or scientific point of such an enterprise would then be opaque to Frege.

Here, right at the origin of analytic philosophy, in a debate between two of the tradition's most influential figures on the nature and purpose of analysis, we find one version of an explicit quarrel between philosophy and poetry—or between, if one prefers, two different conceptions of philosophical (perhaps also poetic) clarity: cognitively expansive (aimed at new truths) and cognitively reflective (aimed at the vivid rearrangement, the reconceiving and recommunication of old truths). An important division of perspectives within and outside the analytic tradition was thus set in motion first in the Frege-Wittgenstein correspondence, and the legacy of this quarrel was formative in the separation of analytic and continental philosophy that was to follow. There is, for example, more than one historical irony in the fact that Heidegger was later to copy the very same Lessing quote into the copy of *Sein und Zeit* he gave to Edmund Husserl in 1927, and write in his own *Holzwege* that "Lessing once said, 'Language can express everything we think clearly"'.[84] For by 1932, invoking the *Tractatus*'s letter and spirit of an "overcoming" (*die Überwindung*) of metaphysics, Carnap would apply to Heidegger's *What is Metaphysics?* more or less the same sorts of criticisms that his teacher Frege had made earlier of the *Tractatus* itself: the demand for sufficiently detailed communication of clear thoughts through the scientific use of a *Begriffsschrift.* One thing the Frege-Wittgenstein correspondence shows is that Wittgenstein never did try to meet that demand, even after Frege's explicit requests. This shows us something important about the conception of philosophy Wittgenstein held, both at the time of the *Tractatus* and afterwards.[85]

[84] Thanks to Wolfgang Kienzler for pointing me toward the Lessing and Heidegger quotations in connection with the Frege letter to Wittgenstein of 16 September 1919 and to Kenneth Haynes, who had pointed me toward the Heidegger quote some years ago, in mind of Wittgenstein (a translation of this quote by Haynes (with J. Young) may be found in *Off the Beaten Track*, trans. and eds. J. Young and K. Haynes (New York: Cambridge University Press), p. 255). Heidegger is said to have copied the Lessing quote into the copy of *Sein und Zeit* that he gave to Edmund Husserl in 1927 (see Husserl, *Psychological and Transcendental Phenomenology and The Confrontation with Heidegger*, trans. and eds. T. Sheehan and R.E. Palmer (Dordrecht: Kluwer, 1997), pp. 21ff). (Daniel Dahlstrom has told me that that Heidegger might have learned of the Lessing source from Paul Lorentz, ed., *Lessings Philosophie: Denkmäler aus der Zeit des Kampfes zwischen Aufklärung und Humanität in der deutschen Geistesbildung* (Leipzig: Meiner Verlag, 1909), p. 98.)

[85] I am grateful for conversations with Enzo De Pellegrin, Norma Goethe, Allan Janik, Wolfgang Kienzler and Brian McGuinness throughout the writing of this

essay, as well as the students in my seminars on Wittgenstein and Frege at Boston University from 2000 to the present who provided me with helpful feedback on the ideas discussed here. A Fulbright research award to Austria gave me time and place to gather primary materials. Burton Dreben, W.V. Quine, G.H. von Wright and participants at the University of California Riverside conference on early analytic philosophy in 1998 (a conference organized through the good offices of Erich Reck) provided helpful encouragement at an early stage in the formulation of my thoughts.

"A Surrogate for the Soul": Wittgenstein and Schoenberg

Eran Guter

One need not be a confirmed Humean in order to observe the effects of habit. When it comes to the contingencies of history, the conjunction of facts and a propensity to relate them to one another might indeed give rise to philosophical confusion. The practice of yoking Ludwig Wittgenstein and Arnold Schoenberg as intellectual comrades-in-arms of sorts seems to have already become commonplace. The *prima facie* appeal of such a practice is undeniable, and, indeed, one could hardly find a text on *Fin-de-Siècle* Vienna that does not underscore at least some similarity between the two great men—their biography, their cultural background, their intellectual projects, their personal fate. In such collage works, historians and philosophers alike often share an enthusiasm for bold brush strokes, which certainly serve a purpose within their overall perspective: to paint a picture of a cultural period to highlight common themes. Yet the thrust of the present essay is, in this sense, antithetical. This is an essay about differences, and some of my brush strokes will be cautious and inevitably tentative. I contend that what sets Wittgenstein and Schoenberg apart from one another is much more interesting philosophically than the historical contingencies that seem to force them together.

My discussion is divided into four parts. I pay a modest tribute in the first section to the historical leads and impasses that serve, so to speak, as a color palette for all those who paint with bold brush strokes. I then move, in the second and third sections, to explicate the various grounds for Wittgenstein's dissenting attitude toward the contemporary music of his time, which I take to be a necessary step in any argument whose conclusion pertains to any relation between the respective ideas of Wittgenstein and Schoenberg. Finally, I turn to Schoenberg's method of composing with twelve tones, framing it in the context of Wittgenstein's philosophical views on music. I shall try to show that the most plausible sense in which Schoenberg's 12-tone system could indeed be rendered a serviceable image for Wittgenstein's view of language is by way of contrast; by underscoring precisely what is unique about Wittgenstein's attitude toward language as music.

E. De Pellegrin (ed.), *Interactive Wittgenstein*, Synthese Library 349, 109
DOI 10.1007/978-1-4020-9909-0_4, © Springer Science+Business Media B.V. 2011

Leads and Impasses

The literature abounds with bold brush strokes. A few major examples should suffice. Hilde Spiel, for instance, is quick to compare the decisiveness with which Schoenberg and his disciples introduced new musical forms that ousted those of the past to the attempts of Wittgenstein and Schlick to purge metaphysics from philosophical thought.[1] William Johnston sketchily suggests that the aphoristic style of Schoenberg's gigantic *Gurrelieder*, his last post-Romantic work, bore an affinity to fragments written by Wittgenstein;[2] and that Wittgenstein (by unmasking self-deception in logicians) and Schoenberg (by deploring excesses of late Romantic music) "unleashed a conservative counterrevolution so drastic as to threaten their own values."[3] Allan Janik and Stephen Toulmin offer an elaborate argument for the relevance of Schoenberg's 12-tone composition technique—peculiarly interpreted as an extension of Karl Kraus's cultural critique into the realm of music—for the understanding of the intellectual milieu from which Wittgenstein's *Tractatus Logico-Philosophicus* emerged.[4]

On occasions, the yoking of Wittgenstein and Schoenberg is merely juxtapositional, suggesting an inert connection via resemblance.[5] However, my concern is with the more ambitious claim that certain technical aspects of Schoenberg's music may be used as a heuristic device for unlocking or shedding light on certain aspects of Wittgenstein's philosophy.[6] Here, I suggest, the terrain is so uncharted that even an experienced traveler might go astray. For instance, in a recent lecture, delivered at Harvard on the occasion of a conference on Schoenberg's chamber music, Stanley Cavell made the following suggestion:

> My suggestion is that the Schoenbergian idea of the row with its unforeseen yet pervasive consequences is a serviceable image of the

[1] Hilde Spiel, *Vienna's Golden Autumn*, 1866–1938 (New York: Weidenfeld and Nicolson, 1987), p. 170.

[2] William M. Johnston, *The Austrian Mind: Intellectual and Social History, 1848–1938* (Berkeley: University of California Press, 1972), p. 139.

[3] Ibid., p. 213.

[4] Allan Janik and Stephen Toulmin, *Wittgenstein's Vienna* (New York: Simon and Schuster, 1973), especially chs. 3 and 8. See also Allan Janik, *Wittgenstein's Vienna Revisited* (New Brunswick, NJ: Transaction Publishers, 2001), ch. 1.

[5] See, e.g., Aldo Gargani, "Techniques Descriptive et Procédures Constructives: Schönberg-Wittgenstein" in J.P. Cometti (ed.), *Ludwig Wittgenstein* (SUD Numéro Hors-série, 1986), pp. 74–121; Friedrich Wallner, "Webern und Wittgenstein: Verbindlichkeit durch Elementarisierung" in Roderick M. Chisholm, Johann Chr. Marek, John T. Blackmore, and Adolf Hübner (eds.), *Philosophy of Mind - Philosophy of Psychology. Proceedings of the 9th International Wittgenstein Symposium* (Wien: Hölder-Pichler-Tempsky, 1985), pp. 482–485.

[6] The reverse case is relatively rare. See, e.g., Wolfgang Hufschmidt, "Sprache und 'Sprachgebrauch' bei Schönberg." *Zeitschrift für Musiktheorie*, 1974, pp. 11–20; James K. Wright, *Schoenberg, Wittgenstein and the Vienna Circle*, 2nd edition (New York: Peter Lang, 2007).

> Wittgensteinian idea of grammar and its elaboration of criteria of judgment, which shadow our expressions and which reveal pervasive yet unforeseen conditions of our existence, specifically in its illumination of our finite standing as one in which there is no complete vision of the possibilities of our understanding—no total revelation as it were—but in which the assumption of each of our assertions and retractions, in its specific manifestations in time and place, is to be worked through, discovering, so to speak, for each case its unconscious row.[7]

What kind of light might Schoenberg's theoretical conception of the 12-tone row throw on Wittgenstein's conception of grammar? Cavell maintains that Schoenberg's use of the 12-tone row exemplifies the communicability of the omnipresence of the inexpressible (or the "unheard," as the title of his lecture suggests)[8]; and, apropos Wittgenstein, such characterization does strike a familiar note, or so it seems. The real question is actually whether the relentless striving for communicability, or rather for comprehensibility—to use Schoenberg's own term[9]—that propels Schoenberg's dodecaphonic compositional procedures is on a par with the relentless, genuinely philosophical striving for the surveyability of grammar. Here, it seems to me, one cannot hope for a real answer before considering seriously what a truly Wittgensteinian response to Schoenberg's work might consist in. Yet such a response is not palpably within reach. It should be stated right at the outset that any attempt to yoke Wittgenstein and Schoenberg for interpretative purposes is bound to occur in a convenient contextual limbo, underplaying a total absence of evidence, of any kind, of any direct influence, interaction or mutual interest between the two men. There is absolutely no reference to Arnold Schoenberg in Wittgenstein's entire *Nachlass* or in the ancillary correspondences that have been made available to scholarship heretofore. Similarly, and perhaps

[7] Stanley Cavell, "Philosophy and the Unheard" in Reinhold Brinkmann and Christoph Wolff (eds.), *Music of My Future: The Schoenberg Quartets and Trio.* Isham Library Papers 5. Harvard Publications in Music 20 (Cambridge, MA: Harvard University Press, 2000), p. 182.

[8] It is noteworthy that Cavell relies here on a 1967 paper by David Lewin on Schoenberg's opera *Moses und Aron.* Lewin's paper was written without the benefit of Schoenberg's so-called "Gedanke manuscripts", which contain the composer's most elaborate attempt to explicate his philosophy of composition. The scholarly edition of these manuscripts appeared only recently in Arnold Schoenberg, *The Musical Idea and the Logic, Technique, and Art of its Presentation*, ed. P. Carpenter and S. Neff (New York: Columbia University Press, 1995).

[9] See, e.g., Arnold Schoenberg, "Twelve-Tone Composition" in *Style and Idea*, ed. L. Stein (Berkeley and Los Angeles: University of California Press, 1975), pp. 207–208.

less surprisingly, there is absolutely no reference to Ludwig Wittgenstein in Schoenberg's literary estate.[10] So here is our first impasse.

Another dead end is the alleged "Labor connection." Composer, pianist and organist, Josef Labor, was "the Wittgenstein family court composer" and musical mentor of some of its members, and for a time he was a well-known musical figure in Vienna. His bust still stands in the garden of the Konzerthaus in Vienna, a forlorn witness to his long forgotten fame. His teaching, composition and musical performances exerted a significant impression on Ludwig Wittgenstein, as we can learn from numerous references in his writings and family letters.[11] He actually counted Labor's music among the very best of Austrian art (MS 107, 184 – CV, 3).[12]

Arnold Schoenberg was also acquainted with Labor. In his autodidactic beginnings as a composer, unsure of his talent and prospects, Schoenberg asked Labor for his opinion on one of his (Schoenberg's) youth compositions. Labor graciously encouraged Schoenberg to pursue a professional career in

[10] For the record, the single appearance of the name Wittgenstein in Schoenberg's literary estate is found in a letter dated November 21, 1913, in which Schoenberg asks his publisher to send a few of his lieder to Frau Bahr-Mildenburg—presumably, Anna Bahr-Mildenburg, the great Austrian soprano—c/o Frau Wittgenstein at Salesianergasse 7, Vienna. Apparently, the reference is to Justine Wittgenstein *née* Hochstetter, wife of Paul Wittgenstein, Ludwig's uncle, who had been residing at this address at the time. See Allan Janik and Hans Veigl, *Wittgenstein in Vienna: A Biographical Excursion through the City and its History* (Wien: Springer-Verlag, 1998), pp. 198–199.

[11] See Martin Alber's comprehensive essay "Josef Labor und die Musik in der Wittgenstein-Familie" in Martin Alber (ed.), *Wittgenstein und die Musik: Ludwig Wittgenstein-Rudolf Koder: Briefwechsel.* Brenner-Studien, vol. 17 (Innsbruck: Haymon Verlag, 2000), pp. 121–137.

[12] I use the following abbreviations for Wittgenstein's standard editions:

BB	*The Blue and Brown Books*
CV	*Culture and Value*
D	*Denkbewegungen: Tagebücher 1930–1932, 1936–1937*
LC	*Lectures and Conversations on Aesthetics, Psychology and Religion*
LW I	*Last Writings on the Philosophy of Psychology, vol. I*
LW II	*Last Writings on the Philosophy of Psychology, vol. II*
NB	*Notebooks 1914–1916*
PG	*Philosophical Grammar*
PR	*Philosophical Remarks*
PT	*Proto-Tractatus*
RPP I	*Remarks on the Philosophy of Psychology, vol. I*
RPP II	*Remarks on the Philosophy of Psychology, vol. II*
TLP	*Tractatus Logico-Philosophicus*
Z	*Zettel*

References to the *Nachlass* are by MS or TS number according to G. H. von Wright's catalogue followed by page number. Translations from the *Nachlass* or from other primary sources in German are my own.

music despite his lack of formal training in music and his lack of proficiency in playing the piano. Years later, Schoenberg expressed his appreciation for Labor's favorable response in a letter sent to the elderly composer, in which he politely acknowledges the gratitude and respect of "modernists" like himself to old masters such as Labor.[13] Schoenberg also included a performance of Labor's clarinet quintet in D major op. 11 in a concert of his Society for Private Musical Performances in Vienna. However, beyond these polite exchanges, and despite Schoenberg's evident familiarity with at least some of Labor's music, there is neither any reference to Labor in Schoenberg's writings on music and musicians nor any reason to believe that Labor had any influence as a composer on Schoenberg's own music. The *New Grove Dictionary of Music and Musicians* (first edition, 1980) seems to have initiated the common misconception that Schoenberg was actually Labor's pupil.[14] In the last analysis, the conjecture of a ready-made musical link between Wittgenstein and Schoenberg through the teaching and influence of Josef Labor remains unwarranted.

However, it is still undeniable that Schoenberg's presence was inescapable in the music scene of central Europe, in particular in Vienna, until he fled the Nazis in 1933 to settle eventually in the United States. As Leon Botstein points out, the kind of outrage expressed at Schoenberg in Vienna during the first decade of the twentieth century surpassed anything that had been leveled against Mahler and Strauss, or even against the new works of Pfitzner, Zemlinsky and Bartók, as Schoenberg drew heavy fire from eminent Viennese critics such as Robert Hirschfeld, Ludwig Karpath and Hans Liebstöckl.[15] It seems unreasonable that his musical activity was totally unknown in the Wittgenstein family, which was so deeply involved and heavily invested in Viennese music; and indeed, quite on the contrary, Karl Wittgenstein, Ludwig's father, actually supported Schoenberg financially at some point.[16] It is also hard to believe that the resounding scandals that occurred in 1907 and

[13] Ernst Hilmar (ed.), *Arnold Schönberg Gedenkausstellung* (Wien: Universal, 1974), p. 160.

[14] See Eric Blom (editorially revised), "Labor, Josef" in *The New Grove Dictionary of Music and Musicians*, ed. S. Sadie (London: Macmillan, 1980), vol. 10, p. 342. This mistake is reproduced in Janik and Veigl, op. cit., p. 124, and was perpetuated for at least the next 20 years by the recently published second edition of *The New Grove Dictionary*. See Eric Blom and Malcolm Miller, "Labor, Josef", *The New Grove Dictionary of Music Online*, ed. L. Macy (Accessed 30 September 2002), *http://www.grovemusic.com*.

[15] See Leon Botstein, "Music and the Critique of Culture: Arnold Schoenberg, Heinrich Schenker, and the Emergence of Modernism in *Fin de Siècle* Vienna" in Juliane Brand and Christopher Hailey (eds.), *Constructive Dissonance: Arnold Schoenberg and the Transformations of Twentieth-Century Culture* (Berkeley: University of California Press, 1997), pp. 3–22.

[16] E. Fred Flindel, "Paul Wittgenstein (1887–1961): Patron and Pianist." *The Music Review*, 32(2), 1971, p. 110.

1913 during major performances of Schoenberg's music in Vienna, the first even involving Gustav Mahler—Vienna's music czar and a distinguished guest in the Wittgenstein Palais—could have escaped the attention of members of the Wittgenstein family.[17] Not unrelated is that fact that Schoenberg's music emerged as a concern in the Wittgenstein family: Paul Wittgenstein, the famous concert pianist, while being no less a nineteenth century man of music than his younger brother Ludwig, made a sincere effort to assimilate the various styles of contemporary music, and yet had no success with Schoenberg's atonal idiom.[18]

A few further contextual observations can also be made. The first decades of the twentieth century proved to be the most dramatic and eruptive period in the history of Western music. According to Paul Griffiths,

> At the moment when the First World War was about to begin, composers from quite different backgrounds, with Debussy, Schoenberg, Stravinsky and Webern at the head of them, had brought about the most rapid and far-reaching changes ever seen in western music. In the course of a few years the standard principles of tonality, formal direction and equilibrium, thematic continuity, rhythmic stability and orchestral homogeneity had all been questioned, sometimes all at once.[19]

It is highly unlikely that Wittgenstein—a probing, well-informed and relentless intellect, immensely sensitive to music—was totally unaware of the violent shock waves emanating from the heart of the European continent, in particular from his native Vienna, which shattered the Western tonal system and threatened to change forever the very essence of music. In fact, Wittgenstein is on record for saying that the music of Alban Berg, Schoenberg's famous pupil and enthusiastic advocate of 12-tone composition, is scandalous.[20] Moreover, David Pinsent noted in his diaries of 1912–1913 vehement arguments between Wittgenstein and his fellow students in Cambridge concerning modern music.[21] On the other hand, we see that when Rudolf Koder reported

---

[17] See Spiel, *Vienna's Golden Autumn*, pp. 171–172.

[18] Flindel, "Paul Wittgenstein (1887–1961): Patron and Pianist," p. 119.

[19] Paul Griffiths, *Modern Music: A Concise History*, revised edition (New York: Thames and Hudson, 1994), p. 50.

[20] Brian McGuinness, *Wittgenstein: A Life. Young Wittgenstein: 1889–1921* (London: Duckworth, 1988), p. 33.

[21] Ray Monk, *Ludwig Wittgenstein: The Duty of Genius* (New York: Free Press, 1990), p. 78. The subject matter of these exchanges remains unknown. Schoenberg's *Five Orchestral Pieces*, op. 16, a thoroughly atonal work, received its première in London in 1912 under the baton of Sir Henry Wood. Whether or not this fact was reflected in any way in these arguments, it is still undeniable that by the time the arguments reported by Pinsent took place, the crisis of the tonal idiom in music, epitomized by Schoenberg's middle period music, was already imminent, and recognizably so, in the high profile works of Richard Strauss and Gustav Mahler.

to Wittgenstein from Vienna about an upcoming high-profile concert of the Vienna Philharmonic Orchestra featuring Bruckner's eighth symphony, he neglected, or found no reason to mention that the evening's program featured also the Viennese première of the three orchestral pieces from Alban Berg's *Lyric Suite*.[22] Such circumstantial evidence suggests that the lack of reference to Schoenberg or to his 12-tone school in Wittgenstein's writings was initially due to a lack of interest rather than to a lack of knowledge. On the eve of the Second World War, when Wittgenstein was in exile in England and Schoenberg in exile in the United States, this lack of interest was probably sealed by a lack of knowledge as well.

The only lead that seems to promise something of an indirect and, as we shall see, ultimately antithetical link between Wittgenstein and Schoenberg is the alleged "Kraus connection." It has been widely acknowledged that Karl Kraus's influential preaching for the purification of language made a long-lasting impression on both men. Even in 1931, after his return to philosophy, Wittgenstein explicitly counted Kraus among the thinkers from whom he took a line of thinking for his own "work of clarification" (MS 154, 33–CV, 19). On his part, Schoenberg gave Kraus a copy of his *Harmonielehre*—a book that contains the germ of his later embrace of atonal idiom in his own compositional practice—with the dedication "I have perhaps learned more from you than one is permitted to learn if one wishes to remain independent."[23] One could also think of a related secondary connection between Wittgenstein and Schoenberg through their respective friendships with Adolf Loos, Karl Kraus's brother-in-arms in the fight against what the two perceived as the culturally malignant aestheticism and hedonism of that time. The facts in this case are established enough. Loos personally supported and promoted Schoenberg's music, and his work clearly inspired certain aspects of Wittgenstein's solemn design of the house in Kundmanngasse in Vienna, built for Margaret Stoneborough-Wittgenstein.[24] It is also clear that both Wittgenstein and Schoenberg admired Loos's work and cultural stance.

Still, the mere acknowledgement of such connections cannot carry us very far in terms of philosophical understanding; and, as it happens, a closer historical look only blurs the big picture. In the case of Loos and Schoenberg, there is at least an apparent asymmetry: for Loos, architecture was not a form of art. Loos argued passionately that while a work of art is revolutionary in its power to tear one out of one's comfortable existence, a house is conservative: it is pleasant, practical, public. Loos's revolt against the upsurge of ornamentation in practical design was, for all present purposes, an attempt

---

[22] Alber (ed.), *Wittgenstein und die Musik*, p. 46.

[23] Quoted in Werner Kraft, *Karl Kraus: Beiträge zum Verständnis seines Werkes* (Salzburg: Müller, 1956), p. 195.

[24] See Paul Wijdeveld, *Ludwig Wittgenstein, Architect* (London : Thames and Hudson, 1994).

to purify language precisely in Kraus's sense of the term. For houses, like the components of language, are artifacts designed for daily use.[25]

In one of his most famous aphorisms, Karl Kraus vividly portrays the cultural mission that he and Loos took upon themselves:

> All that Adolf Loos and I did—he literally, and I linguistically—was to show that there is a difference between an urn and a chamber pot, and that in this difference there is leeway for culture. But the others, the "positive ones," are divided into those who use the urn as a chamber pot and those who use the chamber pot as an urn.[26]

Here we certainly can find more than a merely accidental resemblance in at least Wittgenstein's erstwhile attitude toward language. Now, one might ask, was Schoenberg's quest for "the emancipation of the dissonance" akin to a purification of language in this sense? The answer is both yes and no. On the one hand, as Kraus suggests, the purification of language by means of showing the difference between an urn and a chamber pot entails a corresponding "purification," or rather, liberation of the arts. For the arts must be unbounded by use, the sublime safeguarded. For that reason, Schoenberg of the middle period, the so-called atonal period in his music,[27] enjoyed the critical patronage of Kraus together with Oskar Kokoschka and other artists of the younger generation: the expressionists who dared "to express unmediated a raw and febrile existential truth that honored *no* cultural convention," as Carl Schorske puts it.[28] One is tempted to say that, in defiance of aestheticm, Schoenberg of the middle period presented an urn that could no longer be used as a chamber pot.

On the other hand, Schoenberg's reaction to Post-Romantic excess in music was fundamentally different from Loos's reaction to the Secession movement. The meaning of the emancipation of the dissonance cannot be captured in terms of a sort of removal of a façade of excessively embellished harmony from a *bona fide* musical structure. It is the culmination of a process that was already underway in the music of Gustav Mahler, Richard Strauss, Max Reger,

---

[25] For an illuminating discussion of this comparison between Wittgenstein and Loos see John Hyman, "The urn and the chamber pot" in Richard Allen and Malcom Turvey (eds.), *Wittgenstein, Theory and the Arts* (London: Routledge, 2001), pp. 137–152.

[26] Karl Kraus, *Half Truths and One-and-a-Half-Truths: Selected Aphorisms*, trans. H. Zohn (Chicago: The University of Chicago Press, 1990), p. 69.

[27] Schoenberg never approved of this title, although it has remained in use until the present day. Schoenberg's atonal period extends roughly between 1909 and 1923. It is characterized by an initial outburst of creativity that produced works like *Five Orchestral Pieces*, op. 16, *Erwartung*, op. 17, and *Die Glückliche Hand*, op. 18.

[28] Carl E. Schorske, *Thinking with History: Explorations in the Passage to Modernism* (Princeton, NJ: Princeton University Press, 1999), p. 136.

Claude Debussy, Alexander Scriabin and others, in which the degree of emphasis on non-chordal tones reaches a point where tones lose their inclination to resolve at all. The dissonant harmonic complexes are no longer regulated by an underlying tonal structure but are "set free" as absolute harmonic entities, capable of standing on their own and related solely to one another. In his works from this atonal period, Schoenberg actually offered the arguably inevitable outcome of what he perceived as a complete and irreparable exhaustion of the hierarchic tonal system.[29] Thus, if there is a sense of purification involved in the emancipation of the dissonance at all, it is purification in the sense of stamping out.

Yet by 1923 Schoenberg's expressionist phase reached a dead end. Disillusioned by his prewar, largely non-systematically atonal writing, Schoenberg set himself on a new course toward a rigorously systematized control over the chromatic materials from which he had emancipated himself. As I shall argue below, with this new musical project, Schoenberg decisively, albeit inadvertently, transgressed the Krausian framework of the urn and the chamber pot. Wittgenstein, who never did share Kraus's and Loos's enthusiasm for expressionism in art, and, in particular, for the progressive approach to musical composition, also set himself, before too long, on a new path, disillusioned by his own onetime quest for language in its pure and uncorrupted form, which is to be found underneath the rubble of language as used. Thus, we have reached another historical impasse; the divergent shifts in Wittgenstein's view of language and in Schoenberg's view of the practice of composition circumscribe the grain of truth in the alleged "Kraus connection." Yet this grain of truth is of genuine philosophical importance and in order to pursue it, we must break through this impasse. Hence we now turn to consider Wittgenstein's attitude toward the contemporary music of his time.

Aspects of Decline

Wittgenstein's fierce animosity toward modern music, noted *en passant* in the previous section, is well documented. Yet it is this explicit rejection of modern music that is being patently suppressed when Wittgenstein and Schoenberg are yoked together, rather than serving as a major premise in any attempt to spell out the true nature of whatever relation may obtain between their respective projects.[30] It is worthwhile, I suggest, to look closely at this issue, not simply just as a matter of demarcating Wittgenstein's musical taste, but rather as an important and highly instructive manifestation of his general attitude to his times. And as Georg Henrik von Wright so aptly put it,

> Fichte's famous words "Was für eine Philosophie man wählt, hängt davon ab, was für eine Mensch man ist", may not be interestingly

[29] See Arnold Schoenberg, *Style and Idea*, ed. L. Stein (Berkeley and Los Angeles: University of California Press, 1975), pp. 216–217.

[30] Cf., e.g., Cavell, "Philosophy and the Unheard," p. 177.

> applicable to the average, mediocre, academic philosopher. But for the great ones it is, I think, profoundly true. Their philosophy reflects their personality, and *vice versa*. And if personalities differ profoundly, so will the philosophies. Therefore it is not futile to look for the way in which Wittgenstein's thought can be said to reflect his view of life.[31]

Only three contemporary composers—all of them closely associated with the Wittgenstein family—are actually named in Wittgenstein's writings or in related documentation: Richard Strauss, Gustav Mahler and Josef Labor. As noted before, Labor, the only contemporary composer unequivocally praised by Wittgenstein, is an exceptional case, and I shall have something to say about this later on. The other two composers had already been pressing music hard against the brink of atonality by the first decade of the twentieth century, and, evidently, Wittgenstein was familiar with at least some of their music. Mahler and Strauss each had a crucial role in driving Western tonal music into a dead end that resulted in nothing short of a crisis in musical language itself. This fact has been observed by Felix Salzer,[32] and the particular reference to this eminent musicologist in the present context is, of course, far from being accidental. Felix Salzer was Ludwig Wittgenstein's nephew, and according to Brian McGuinness, the two men spent some time together discussing Salzer's own work and the music theory of Heinrich Schenker, who was Salzer's mentor.[33] These discussions began in 1926 and continued on to summers on the *Hochreit*, the Wittgenstein family country estate, in the early 1930s.[34]

These intellectual exchanges on music set up an important nexus of ideas for our discussion. Evidently, Wittgenstein and Salzer shared an overall pessimism with regard to the prospects of recent musical innovations. This brand of cultural pessimism is clearly traceable to Oswald Spengler, on the one hand, and to Heinrich Schenker, on the other. Wittgenstein came under the spell of both thinkers around the same time. While exchanging ideas on music with Salzer, he was also reading Spengler's magnum opus, *The Decline of the West*, in the late spring of 1930. Both thinkers enjoyed at least some credit in his eyes. According to Salzer, Wittgenstein's judgement of Schenker's view of mu-

[31] Georg Henrik von Wright, *The Tree of Knowledge and Other Essays* (Leiden: E. J. Brill, 1993), p. 90.

[32] Felix Salzer, *Structural Hearing: Tonal Coherence in Music*, 2 vols. (New York: Dover Publications, 1952), pp. 5–6.

[33] Salzer remained a champion of Schenker's theories all his life. At various times he edited two journals, first *Der Dreiklang* and later *The Music Forum*, which were dedicated primarily to the study of Schenker's theories. His famous pedagogic textbook *Structural Hearing* (op. cit.) is an attempt, rendered quite successful by many, to enhance Schenker's ideas and methods and rework them into a systematic course of study.

[34] Felix Salzer reported this to Brian McGuinness. I am grateful to Professor McGuinness for relaying this information to me (personal communication, 3/1/2002).

sic was not entirely negative.[35] As for Spengler, despite being critical about what he perceived to be a number of irresponsible ideas in *The Decline of the West*, Wittgenstein nevertheless wrote in his diary on May 6, 1930: "Many, perhaps most [of Spengler's ideas] are in total accordance with what I have been thinking myself." (D, 24) Of course, the mere conjunction of these facts does not imply that Wittgenstein was inclined to entertain the ideas of the two thinkers on the same track. Yet, luckily, we do have a "smoking gun": by 1931 Wittgenstein felt himself versed enough in Schenker's approach to music to relate it to his own notion of "family resemblance," which he had adapted from Spengler's morphological approach to cultural epochs.[36]

The direct influence of Oswald Spengler, a philosophical dilettante full of sound and fury, on Wittgenstein's work—corroborated by the latter's own admission (MS 154, 33 - CV, 19)—caught most scholars by surprise in 1977 upon the publication of the posthumous volume *Vermischte Bemerkungen* (later published in English as *Culture and Value*). However, since then, this influence has been widely acknowledged, and it is now fairly established that Wittgenstein espoused Spengler's views on two major fronts. First, as Georg Henrik von Wright has argued on various occasions, Wittgenstein shared Spengler's cultural pessimism and his perspective of epochal decline. According to von Wright,

> Wittgenstein did not, like Spengler, develop a philosophy of history. But he lived the "*Untergang des Abendlandes*", the decline of the West, one could say. He lived it, not only in his disgust for contemporary Western civilization, but also in his deep awe and understanding of this civilization's great past.[37]

Furthermore, according to Rudolf Haller,

> Wittgenstein finds in Spengler not only an intellectual kinsman, who declares his alienation from the surrounding civilization, with its symptoms of a declining epoch, but also the initiator of an approach or "line of thinking" which seems to him most appropriate as the methodological tool for the investigation of language games.[38]

This "line of thinking" is the main principle of comparative morphology or the "physiognomic method," originally derived from Goethe's writings—a

[35] Brian McGuinness, personal communication, 2/16/2000. I will have more to say about Wittgenstein's attitude toward Schenker below.

[36] This striking reference appears in the form of a handwritten comment—"Schenkersche Betrachtungsweise der Musik" (TS 213, 259v)—on the occasion of introducing the concept of "family resemblance."

[37] Georg Henrik von Wright, "Ludwig Wittgenstein in Relation to his Times" in Brian McGuinness (ed.), *Wittgenstein and his Times* (Chicago: The University of Chicago Press, 1982), p. 116.

[38] Rudolf Haller, *Questions on Wittgenstein* (Lincoln, Nebraska: University of Nebraska Press, 1988), p. 80.

conceptual iceberg, of which Wittgenstein's adaptation of the pervasive notion of "family resemblance" in his later writings is merely the tip.

Wittgenstein's famous 1930 sketch for a foreword to his *Philosophical Remarks* provides a sweeping impression of Wittgenstein's alienation from the contemporary art of his time (he names modern music and architecture in particular) and its deceptive spirit of progress (MS 109, 204ff. – CV, 6–7). The "great suspicion (though without understanding its language)" with which Wittgenstein approached modern music, by his own admission, and his lamentation of "the disappearance of the arts" mark a clear point of convergence with Spengler's somewhat more furious yet strikingly similar remarks on the impotence and falsehood of contemporary art:

> What do we posses today as "art"? A faked music, filled with artificial noisiness of massed instruments; a faked painting, filled with idiotic, exotic and showcard effects, that every ten years or so concocts out of the form-wealth of millennia some new "style" which is in fact no style at all since everyone does as he pleases.[39]

Spengler's fingerprints are unmistakable also in Wittgenstein's later comments on the deterioration of high culture in his 1938 lectures on aesthetics, especially in Wittgenstein's characterization of artistic decline in terms of a breakdown of artistic necessity through reproduction of artifacts and a corresponding deterioration of sensitivity leading to indifference, and also in his curious remark concerning vintage furniture (LC, 7).[40]

So much is obvious; yet I suggest that Spengler's impact on Wittgenstein's thinking about art runs deeper still. To realize this, we need to turn now to Heinrich Schenker. Schenker's pessimism concerning the prospects of modern music is intrinsically related to his unique view of musical composition. Working up his case by meticulously analyzing masterworks of Western music, Schenker theorized that works of music that are tonal and exhibit mastery are temporal projections of a single element: the tonic triad. According to Schenker, the projection of this triad comprises two processes: (a) the transformation of the triad into a basic contrapuntal design, which he called *Ursatz*[41]; and (b) the *Auskomponierung*, or elaboration, of the *Ursatz* by various techniques of prolongation. This notion of music is highly abstract; in practice, as Schenker shows in his own analyses, the process of elaboration begins when the *Ursatz* is already in an articulated form—this he called *Hintergrund*, or the "background", of the work. The number of possible forms of background is theoretically infinite.

---

[39] Oswald Spengler, *The Decline of the West*, trans. C.F. Atkinson (New York: Alfred A. Knopf, 1939), vol. 1, p. 194.

[40] Cf. ibid., pp. 293–295.

[41] The *Ursatz* is made of a fundamental line, or *Urlinie*, which is a linear descent to the root of the triad. The *Urlinie* is accompanied by an "arpeggiation" in the bass (*Bassbrechung*) from the tonic to the dominant and back.

Yet, at the heart of Schenker's abstract notion of music, one finds the conviction that the masterworks of Western music teach us that hearing music consists in recognizing a structural standard, which is shared by anything that we may rightfully call music. Hearing music as an exfoliation of this fundamental structure is part of the "phenomenology" of musical perception, rather than a matter of inference or analysis. As Milton Babbitt pointed out, the crucial idea in Schenker's view of music is "the perception of a musical work as a dynamic totality, not as a succession of moments or a juxtaposition of 'formal' areas related or contrasted merely by the fact of thematic or harmonic similarity or dissimilarity."[42] According to Schenker, all works of music (in particular all masterworks) are, in a sense, extended commentaries on the tonic triad. In effect, Schenker's theory embodies an attempt to describe musical thinking itself: it describes how we keep a single triad in mind over a period of time, and how we interpret configurations of notes as contributing to the continuity of that cognition.

Thus, it becomes a matter of analytic truth that all works of music that digress from triadic tonality—that is, whose Schenkerian analysis shows that their surface, or "foreground", cannot be hierarchically related by a series of expansions ("middleground" layers) to a constant "background", and ultimately, to the *Ursatz*—are to be patently rejected by Schenker as unsuccessful, superficial, or altogether musically nonsensical, depending on how severe the digression is. Schenker's hostility toward contemporary music was fueled not only by his mighty theory of music, but also, and perhaps even more significantly, by his conviction that the results of his theory betoken a disintegration of musical culture on all fronts.[43] Irreverence toward the laws of tonal effect, he believed, reflects a loss of musical instinct for the inner complexities of the masterworks of Western music among performers and composers alike, which in turn hinders the musician's almost sacred mission to provide access to the world of human experience contained in such masterworks. Thus, he likened contemporary music making to a Chinese person picking up a text by Goethe without having sufficient knowledge of the German language.[44]

In the face of the dramatic changes in compositional techniques that had taken place at the turn of the twentieth century, Schenker stated as early as 1910 that music, like the once-great cities of Herculaneum and Pompeii, lay in ruins. He openly disapproved of the compositional practice of Mahler,

---

[42] Milton Babbitt, "Review of Structural Hearing by Felix Salzer," *Journal of the American Musicological Society*, 5, 1952, p. 262.

[43] See Robert Snarrenberg, *Schenker's Interpretative Practice* (Cambridge: Cambridge University Press, 1997), pp. 145–150. I am indebted here to Snarrenberg's useful survey of a variety of primary sources.

[44] In his 1938 lectures on aesthetics, Wittgenstein uses a similar image of someone who admires a sonnet admitted to be good without knowing English (LC, 6). Again, for Wittgenstein, music exemplifies the point being made: there is an intimate link between artistic experience of art and what he calls *Menschenkenntnis*, our knowledge of human nature (cf. PI II xi 227).

Strauss, Reger, and Schoenberg.[45] He also deplored the fact that people no longer made distinctions between the output of composers like Debussy, Ravel and Stravinsky, and the masterworks of Bach, Mozart and Beethoven, and, in effect, treated them as if they were all "music" in the same sense. In Schenker's view, the emancipators of the dissonance were merely reveling in empty sonorities, being unable to bind them together as elaborations of a single chord.[46] Of course, Arnold Schoenberg's musical and theoretical output was an anathema for Schenker, and the two men were entangled in bitter, extensive polemics against one another.[47] I shall return to the roots of this dispute in the next section.

My précis of Schenker's highly technical writings is inevitably oversimplified. Yet it allows us to see fairly easily how the main thrust of Schenker's view of music coincides with Wittgenstein's thinking. In section 58 of the so-called "Big Typescript" (cf. MS 111, 119 – CV, 14), Wittgenstein reprimands Spengler's dogmatism in sorting cultural epochs into families, ascribing properties, which only the prototype, or archetype (*Urbild*) possesses, to the object that is viewed in its light. This is the context in which Wittgenstein saw a connection with the Schenkerian view of music. The Schenkerian *Ursatz*, which encapsulates the whole of triadic tonality, is the *Urbild* in Wittgenstein's analogous construal. Hence, analogously, Schenker's mistake is in the way that he extends the scope of statements true of tonality (in its pre-articulated form) to particular works of tonal music. Clearly, behind this mistake stands the "craving for generality" that Wittgenstein often diagnoses and condemns (see, e.g., BB, 17–18). Indeed, Schenker seems to have envisioned that his theory amounts to nothing less than a fully-fledged essentialist account of music, a complete analytic definition of the concept of music, which lays down necessary and sufficient conditions for its application, and hence, as we have seen

45 At some point Schenker wanted to publish an inflammatory essay titled "On the Decline of Compositional Art: A Technical-Critical Investigation." His publisher, Emil Hertzka, who was also the music publisher of Mahler, Strauss and Schoenberg, undermined this project.

46 For instance, he accused Richard Strauss of trying to mask the primitive design of his music with heavy orchestration, with noise and polyphonic clatter, and of resorting to vulgar, extra-musical narratives in order to solve problems of musical continuity. As for Max Reger's music, Schenker's attempt to analyze Reger's quintet op. 64 suggested to him that the celebrated German composer had been abandoned by all instincts for music. It is noteworthy that Spengler expressed a similar opinion of the totally aloof character of Reger's music: "In the real command of a language there is a danger that the relation between the means and the meaning may be made into a new means. There arises an intellectual art of *playing* with expression, practiced by ... Reger in music" (Spengler, *The Decline of the West*, vol. 2, pp. 136–137).

47 These disputes can be seen most openly in Schoenberg's *Harmonielehre* (1910) and in the second volume of Schenker's *Die Meisterwerke in der Musik* (1926).

before, entails a clear demarcation between *bona fide* cases of music and what is to be regarded, in lieu of a better term, as non-music.

The upshot of Wittgenstein's conflation of Schenker and Spengler is this. Wittgenstein is committed to the contention that triadic tonality is the focal point for comparing musical instances; he also maintains that various musical instances may bear more or less family resemblance to one another, to the extent of the exclusion of certain instances. Yet Wittgenstein is bound to deny that the general validity of the concept of tonality depends on the claim that everything which is true only of the abstract Schenkerian *Ursatz* (*qua* the prototype of the observation) holds too for all the musical instances under consideration. Rather, when the prototype is clearly presented for what it really is, and thus becomes the focal point of the observation, the general validity of the concept of tonality will depend on the fact that it characterizes the whole of the observation and determines its form. In this anti-essentialist vein, the Schenkerian *Ursatz* becomes a mere methodic device that can be laid alongside the musical instances under consideration as a measure.[48]

While Wittgenstein never did address the concept of tonality directly, at least the rudiments of what we might call his "philosophical conception of tonality" can still be extracted from what he did write about such matters as the rules of harmony and their effects. I shall dedicate the remainder of this section to the fleshing out of this crucial issue. Yet, before I do that, a few general remarks on Wittgenstein's various texts on musical experience are in order. The bulk of these texts belong, by and large, to his later work, and they are thematically indigenous to his thinking on philosophical psychology. Wittgenstein's discussion of musical experience occurs at the intersection of three often-overlapping concerns: (a) the grammatical complexity of language games that pertain to aesthetic phenomena and to musical experience in particular; (b) the pervasiveness of aspect dawning, in particular in music; and (c) the notion of physiognomy and its philosophical ramifications. I have dealt with these issues in some detail elsewhere,[49] so I will restrict myself here to a brief summary.

According to Wittgenstein, our intercourse with music exemplifies a special kind of grammatical complexity: each move in the language game played logically presupposes corresponding moves in various other games, ultimately presupposing "the whole range of our language games" (MS 132, 59 – CV, 52). In this sense of a logical hierarchy between language games, we may speak

[48] Thus it becomes quite clear in what sense Wittgenstein's judgement of Schenker's view of music must have been forthcoming to an extent, as Felix Salzer reported, and perhaps even why he also told Salzer (concerning the latter's own rendition of Schenker's theory) that he hopes that Salzer "has boiled it down" (reported by Brian McGuinness, personal communication 2/16/2000).

[49] See Eran Guter, "Wittgenstein on Musical Experience and Knowledge" in Johann C. Marek and Maria E. Reicher (eds.), *Experience and Analysis, Contributions to the 27th International Wittgenstein Symposium* (Kirchberg am Wechsel: Austrian Ludwig Wittgenstein Society, 2004), pp. 128–130.

of the meaning of a musical gesture in terms of what I would like to call a "vertical axis".[50] Yet what makes our musical experiences akin to aspect dawning on the one hand, and to what Wittgenstein calls "the use of words in a secondary sense" (PI II xi 216), on the other, is the manner in which we reach for a certain expression as the only possible way in which to express our experience, our perceptions, inclinations and feelings. Just like our expression of aspect dawning or our use of words in a secondary sense, the specificity of a musical gesture lies in the absence of a "more direct" way of expressing the experience in question. Here, according to Wittgenstein, music and language intertwine, or as he put it, "the [musical] theme is in reciprocal action with language" (MS 132, 59 – CV, 52); the relation between the musical gesture and the thing expressed is internal.[51]

An important facet of Wittgenstein's discussion of musical expression is the logical implications of his emphasis on the notion of physiognomy. The notion of physiognomy—the meaningful irregularity of the living body—is central both to his explication of aspect seeing (PI II xi 193) and to his various discussions of musical expression (LC, 4; PI §536; RPP I §434; CV, 52). According to Wittgenstein, enormous variability, irregularity, and unpredictability are an essential part of human physiognomy and the concepts for which human physiognomy serve as a basis (RPP II §§614–615, 617, 627). Musical gesture is akin to human physiognomy in being fundamentally non-mechanical; it cannot be recognized or described by means of rules, and it introduces indefiniteness, a certain insufficiency of evidence, into our musical understanding that is constitutive in a logical sense, hence not indicative of any deficiency of knowledge (see, e.g., MS 137, 67 – CV, 73; RPP II §695; Z §157). The concept of musical expression, like the concept of "soul", is diametrically opposed to the concept of a mechanism (cf. RPP I §324)—an exact, definite calculation and prediction is conceptually detrimental to what we regard as musical expression. Thus, musical gesture admits what Wittgenstein calls "imponderable evidence": "subtleties of glance, of gesture, of tone" that

[50] The term "vertical" in this context is adopted from Michel Ter Hark, *Beyond the Inner and the Outer: Wittgenstein's Philosophy of Psychology*. Synthese Library, vol. 214 (Dordrecht: Kluwer, 1990), pp. 33–42.

[51] By "internal relation", Wittgenstein means a relation that is given with, or at least partly constitutive of, the terms adjoined. Such a relation cannot be established by examining the relata, since we could not identify the relata independently. The relata are adjoined in practice, so their relation is effected by the way we identify things. Thus, an internal relation is to be found in grammar. Wittgenstein's great insight was that musical meaning is an internal relation, or a grammatical relation, not a relation between music and something else. Indeed, as Roger Scruton correctly observed in a recent paper, "analytical philosophy of music has grown around the question of musical meaning, which became articulated, during the twentieth century, in ways that were inimical to Wittgenstein's vision" (Roger Scruton, "Wittgenstein and the Understanding of Music", *The British Journal of Aesthetics*, 44(1), 2004, p. 1).

form a basis for our *Menschenkenntnis*, our "knowledge of mankind"—a kind of knowledge or a skill that can be learned by some and taught by some, yet only through "experience" or "varied observation" and by exchanging "tips" (PI II xi 227–229). The imponderability of this kind of evidence is significantly reflected in the way we attempt to express our experiences, and in the measure of the success of what we offer as our "justifications"; that is, significantly, in our interlocutor's willingness to follow the rules of the game that we are playing, using concepts based on indefinite evidence (LW I §927).

Considering musical meaning, an internal relation adjoining musical gesture and the life of mankind (shown by language), enables us to appreciate Wittgenstein's assertion: "For me this musical phrase is a gesture. It insinuates itself into my life. I adopt it as my own." (MS 137, 67 – CV, 73) What we nonchalantly call "music" or "a melody" is already given to us with a familiar physiognomy, its impression vertically related to a myriad of other language games in its significantly human environment. And so, says Wittgenstein, "understanding music is an avowal of the life of mankind." (MS 137, 20) Wittgenstein's contention that music opens up a realm of *Menschenkenntnis* for us to partake underscores the strong affinity with Heinrich Schenker's aforementioned view of the role and the profundity of the great masterworks of Western music in providing access to the world of human experience.

Let us now turn to Wittgenstein's treatment of the notion of *Harmonielehre*. It should not be surprising that Wittgenstein regarded *Harmonielehre*, the systematic representation of the rules of tonal effect, as grammar.[52] *Harmonielehre* typically describes the way we hear harmonic relations and prescribes methods for constructing chord progressions in a way that renders these relations clear and distinct. As one would expect, Wittgenstein maintains that a musical passage is not an arbitrary string of sounds; the right way to combine a musical tone with other tones is somehow already built into the tone itself.[53] In fact, this was considered the essence of tonality from roughly 1600 to 1910: a mere sequence of notes is not a musical phrase until it is heard as organized around one privileged tone, namely, the tonic. Wittgenstein says in acknowledgement:

> [t]he finitude of the musical scale can only derive from its internal properties. For instance, from our being able to tell from a note *itself* that it is the final one, and so that this last note, or the last notes,

[52] It is noteworthy that, traditionally, the term *Harmonielehre* does not denote primarily a kind of abstract treatise on the nature of harmony, but rather refers to a practical handbook designed to teach a beginning pupil how to become a composer of tonal music through an explanation of rules and their application, accompanied by a standard regimen of examples and exercises.

[53] Wittgenstein stated this idea explicitly as early as 1915: "Nor is a melody a mixture of tones, as all unmusical people think." (NB, 41; cf. PT 3.1602 and TLP 3.141)

> exhibit inner properties which the notes in between don't have. (PR §223)

The privileged status of the tonic is a property which it cannot fail to possess, because it is essential to its being the thing it is. Thus, according to Wittgenstein, tonal relationships, represented by *Harmonielehre*, are internal, that is, they cannot fail to obtain, since they are given with, or constitutive of, the relata in practice; they cannot be underpinned or explained by postulating mediating links between the relata.

Wittgenstein illustrated this point in *The Brown Book* by discussing the phenomenon of hearing the same tone again in a diatonic scale (BB, 140–141)—certainly one of the most fundamental tonal effects in music. The question is why we call tones that appear in an interval of an octave "the same tone". Wittgenstein asks us to imagine a case in which a person calls the tonic, the dominant and the octave, "the same tone". His point is that we can say that this person hears different things than we do, insofar as we do not assert that there must be some other difference between this person and us besides the one mentioned before. Simply put, the only thing that we can say in this case, and in any similar case, is that two tones that stand in the relation of "sameness" cannot be a tonic and a dominant, or a dominant and an octave. Tonality is mirrored by grammatical analysis.[54] Thus *Harmonielehre* represents the grammar of tones in a way that is analogous to Wittgenstein onetime example of the color octahedron (see PR §1, §3): it is "at least in part phenomenology and therefore grammar." (PR §4)

We can see that quite in accordance with Wittgenstein's general view of musical meaning, which I described above, the most important feature of his treatment of *Harmonielehre* is the emphasis on its being a representation of internal relations, hence on the primacy of praxis. Tonality (experiencing and expressing certain relationships between tones) is effected by the way we recognize and describe things, and to that extent *Harmonielehre* is not a matter of taste (PR §4). In 1934, Wittgenstein wrote:

> Is the *Harmonielehre* constructed in accordance with our feelings; do we try out whether a [chord] progression pleases us [insertion: more or less], in the way that we perhaps select the ingredients of a dish according to our taste? And is the difference perhaps in that there are valid rules for the taste in chord progressions that are more general than [the rules for the taste] in food? Could *one* reason be given at all for why the *Harmonielehre* is the way it is? And, first and foremost, *must* such a reason be given? It is here and it is part of our entire life. (MS 157a, 24–26)

The comparison between rules of grammar and rules of cookery is standard in Wittgenstein's later writings (see, e.g., PG §133; TS 213, 236). It serves the

[54] Hypothetical objects, such as the vibrations of the air, the miniscule apparatus of the inner ear etc. are patently excluded (cf. PR §218).

purpose of highlighting the unique status of rules of grammar. According to Wittgenstein, rules of cookery can be justified by appealing to their (external) purpose, i.e., creating a delectable dish. The goal of cooking is independent of the rules of cookery: if I decide to improvise in the kitchen, I should bear in mind the old maxim that the proof of the pudding is in the eating. The rules of cookery are therefore constituted externally by the desired result, which is causally effected by the activity of cooking and by various other contingencies (such as the kind of ingredients that I happen to have in my kitchen cabinet etc.). Thus we may legitimately speak of right or wrong rules of cookery (to wit, those which happen to bring about a tasty dish are right). In contradistinction, if rules of grammar define a practice (i.e., if they do not admit alternatives), then they cannot be said to be constituted externally in this sense. A systematic deviation from the rules of grammar entails a wholesale rejection of the practice defined by those rules.

The status of *Harmonielehere* as a representation of grammar means that, for Wittgenstein, tonality sets limits to what makes (musical) sense. Here Wittgenstein seems to be in complete agreement with Heinrich Schenker. Still the profundity of Wittgenstein's philosophical emulation of Schenker's view of music is revealed when we consider that for Wittgenstein, what a musical gesture means is determined by its "vertical axis", that is, by consisting in a move in a vertically complex language game of the kind described above. Wittgenstein makes this explicit in the following passage from 1946:

> We can apply to the melodies by the various composers the principle: each species of tree is a "tree" in a different *sense* of the word. That is, don't be misled by the fact that we say all these are melodies. They are stages along a path which leads from something you would not call a melody to something else that you would equally not call a melody. If we just look at the sequences of notes and changes of key all these entities [*Gebilde*] seem to be in coordination [*in Koordination*]. But if you look at the environment in which they exit [*das Feld in dem sie stehen*] (and hence at their meaning), you will be inclined to say: In this case melody is something quite different from what is in that one (amongst other things, here it has a different origin and plays a different role). (MS 131, 12 – CV, 47)[55]

From Wittgenstein's perspective, we can say that by showing that (great) works of music are, in the last analysis, extended commentaries on the tonic triad, Schenker has merely given us a focal point or a measure for the observation that each instance of a musical gesture is a gesture in a different sense of the word.[56] Wittgenstein's somewhat cryptic way of defining musical gesture

[55] I modified Peter Winch's translation.

[56] Regrettably, as it happens in many of Wittgenstein's remarks on music, his non-standard use of technical terms in music (e.g., "melody") often results in the obfuscation of his philosophical point.

as "stages along a path" that adjoins what is *not yet* music with what is *no more* music, betokens of an internal relation, which, as I have suggested, is the mark of musical meaning. Hence, for Wittgenstein, looking at the meaning of a musical gesture amounts to looking at the actual language game in which it is embedded, and its vertical relation to a range of other language games.

Since tonality cannot be vindicated by reference to putative facts about the world or about the mind, as Schenker believed, there is no sense in seeking the reason why *Harmonielehre*, the grammatical representation of tonal manifestations, is the way it is. As Wittgenstein put it, "[that reason] is here and it is part of our entire life"; that is, the musical distinctions that we make have to be important to us, given the kind of beings we are, the purposes we have, our shared discriminatory capacities, and certain general features of the world we inhabit. This leaves more than ample room for composers to extend the range of musical expression (see LC, 6; MS 133, 30 – CV, 55). Yet the boundaries of sense are also clear, and they suggest two important angles on the decline of modern music. First, since the rules of harmony are not constituted externally by concocting chord progressions according to taste, those composers who do so, those who revel in empty sonorities (to use Schenker's phrase), tarnish the tonal idiom from within, so to speak, by ungrammatical and effectively senseless gesticulation.[57] Still, a comprehensive *Harmonielehre*—"just looking at the sequences of notes and changes of key"—can readily expose such grammatical mishaps for what they are: simply wrong. Second, a much more serious transgression of tonality would amount to a wholesale rejection of its praxeological foundation, ultimately a nonsensical transgression of the "reason" why the practice is the way it is, to wit, "our entire life." Wittgenstein's point is that such a perversion of musical gesture could no longer be either right or wrong, for it would amount to "speaking of something else" (cf. PG §133). To this latter, deeper worry we shall now turn.

The Music of the Future

It would be worthwhile, I think, to take a closer look at some of Wittgenstein's remarks on the music of Gustav Mahler—praised already in his lifetime as "the contemporary of the future"—the only truly modern composer, who apparently was great enough in Wittgenstein's eyes to be worthy of attention. Wittgenstein's somewhat abusive remarks on Mahler, those scattered in his various writings and those relayed to us by friends and disciples, exemplify a distinct duality toward Mahler's musical *persona* that was typical

[57] Indeed, as Robert Morgan points out, we can see that even in Richard Strauss's most progressive music, such as his operas *Salome* (1905) and *Elektra* (1908), while stretched to its outermost limits, tonality is still present as an underlying control. See Robert P. Morgan, *Twentieth-Century Music: A History of Musical Style in Modern Europe and America* (New York: W. W. Norton, 1991), pp. 32–33.

among Austrian *literati* at that time. Carl Schorske describes this as a duality in Mahler's functional relation to the classical tradition; an acute tension between Mahler's acceptance as a conductor—a guardian of the abstract, autonomous music so cherished by the educated elite—and his rejection as a composer, who subversively attempted to imbue abstract high-culture music with concrete vernacular substance.[58] Georg Henrik von Wright recalled from his conversations with Wittgenstein that Wittgenstein had a tremendous respect toward Mahler, not only as a conductor (Wittgenstein thought that as a conductor, Mahler was unequalled), but also as a composer, although he did believe strongly that there is something deeply faulty in Mahler's music.[59] Mahler was a genuine problem for Wittgenstein, a limiting case in the history of Western music—"You would need to know a good deal about music, its history and development, to understand him," said Wittgenstein at one point.[60]

Let us first consider two passages that Wittgenstein wrote in 1931:

> When the later ones of the great composers *sometimes* write in simple [variant: clear] harmonic progressions [variant: relations], then they bear witness to their ancestral mother [*Stammutter*]. Mahler appears to me precisely at these moments (when the *others* move me the most) exceptionally unbearable, and I always would like to say: but you merely heard this from the others, this does not (*really*) belong to you. (D, 47)

> A picture of a complete apple tree, however accurate, is in a certain sense much less like the tree itself than is a little daisy. And in the same sense a symphony by Bruckner is infinitely closer to a symphony from the heroic period than is one by Mahler. If the latter is a work of art it is one of a *totally* different sort. (But this is actually itself a Spenglerian observation.) (MS 154, 39 – CV, 20)

Given Wittgenstein's contention that musical meaning is an internal relation in which music and language are in reciprocal action, it is clear why he says that he is moved most strongly when composers (other than Mahler) write in the clearest tonal idiom. Such familiar musical gestures—a perfect cadence, for instance—are already deeply entrenched as parts of our life; as Wittgenstein put it, what is ordinary is filled with significance (cf. MS 132, 59 – CV, 52). Yet Wittgenstein believes that precisely in such moments of great expressive transparency, a transgression is taking place in Mahler's music: the avowal is not genuine. It is crucial to understand exactly why Wittgenstein finds Mahler

---

[58] Carl E. Schorske, *Thinking with History*, pp. 172–174.

[59] Reported by Enzo De Pellegrin from an interview with G.H. von Wright, which took place in Helsinki, Finland in early summer of 1999. I am grateful to Dr. De Pellegrin for relaying to me relevant segments from this conversation.

[60] Quoted in John King, "Recollections of Wittgenstein" in Rush Rhees (ed.), *Recollections of Wittgenstein* (Oxford: Oxford University Press, 1984), p. 71.

most unbearable when he writes in simple tonal relationships, rather than in his many moments of extraordinary harmonic daring. Perhaps the best way to approach this problem is to acknowledge Wittgenstein's striking grasp of the essence of Mahler's musical language. Mahler's mature works—e.g., his fourth symphony—display significant ambivalence in the area of harmony and tonal relationships. On the one hand, the music often seems deceptively conservative, employing undisguised dominant relationships that still play an essential structural role. On the other hand, as Robert Morgan observes, "tonality in Mahler comes close to reaching its final stage of dissolution: complete works, and even individual movements, no longer necessarily define a single key, but explore a range of related and interconnected regions, often closing in a different key from the one in which they began. ... Such procedures alter the very meaning of tonality, which becomes a complex network of interchangeable relationships, rather than a closed system that ultimately pulls in a single, uncontested direction."[61]

This observation suggests that Wittgenstein's criticism of Mahler's music focused on its allegedly perverse toiling with tonality rather than on the vagaries of over-stretched chromaticism. In other words, Mahler was a problem for Wittgenstein because his musical gestures only play at being in reciprocal action with language (and our life), and what is familiar cannot be otherworldly at the same time. Wittgenstein's comparison between Mahler and Bruckner provides further support of this view. To a significant extent, the music of these two great composers exhibits strikingly similar surface characteristics: the evident employment of Wagnerian compositional techniques, extended chromaticism, the enormous length of their works that extends far beyond the traditional symphonic form, the juxtaposition of contrasting musical materials, etc. Interestingly, Wittgenstein's reaction to such musical innovations was quite favorable, as we can learn, for example, from his enthusiastic correspondence with his sister Hermine concerning Bruckner's third symphony in D minor, in particular its third movement, the *scherzo*.[62]

In a letter dated January 22, 1948 to Ben Richards, Wittgenstein's commented on this work:

> *Of course* what you say about the ending of the 3rd movement (Bruckner) isn't final. His "abruptness" is an essential part of his language. He writes in "main clauses" (I'm not sure if that's the right grammatical term; I mean the opposite of "subordinate clause"). He doesn't

[61] Morgan, *Twentieth-Century Music*, p. 22.

[62] See Brian McGuinness, Maria Concetta Ascher and Otto Pfersmann (eds.), *Wittgenstein Familienbriefe. Schriftenreihe der Wittgenstein-Gesellschaft,* vol. 23 (Wien: Verlag Hölder-Pichler-Tempsky, 1996), pp. 131–133. It is noteworthy that this symphony was the first to exhibit Bruckner's mature style, which was greeted with fierce hostility by the Viennese audience and critics in its premiere in 1877—Brahms condemned Bruckner's works as being a *symphonische Riesenschlange.* I may also add, as an anecdote, that young Gustav Mahler was one of the very few people in the audience who stood up applauding at the end of the performance.

> say "If it rains I shan't go", but "It rains. I don't go." A good example of it is the introduction to the first movement, which sounds like so many scraps but is a connected whole. People generally, when they first hear Bruckner, and for a *long* time, can't hear his music "*connected*". In the same way that ending of the 3rd movement is *not* abrupt, but of course it seems so, unless you can listen to his way of telling the story. (By the way, the 3rd movement does not lead into the 4th.)[63]

What Wittgenstein calls "Bruckner's way of telling the story" pertains essentially to Bruckner's typical approach to the large-scale tonal-narrative of his symphonies. As Wittgenstein suggests, this is intrinsically related to the problem of hearing Bruckner's music as a "connected whole". The first obstacle for anyone who tries to hear Bruckner's music "connected" is its unprecedented monumentality. The sheer size of a Bruckner symphony is attained mainly through a slowing of usual musical processes. The ideal of the Classical sonata form—which underlies the symphonies of Haydn, Mozart and Beethoven, for example—was dynamism: the music evolved both tonally and motivically to create the effect of goal-oriented forward motion. This dynamism was created both by harmonic motion and logical motivic transformation. Yet, as seen clearly in the first movement of Bruckner's third symphony, instead of the classicist's brief contrasted themes, skillfully bridged by interludes, Bruckner upsets the sonata form by setting forth a number of independent theme-groups, each consisting of well-contrasted motivated portions, and allowing each motif its full elaboration and expanse of time. This results in huge stretches of thematic development, and expansive sections of static harmony.[64]

Bruckner's unaccustomed juxtaposition of blocks of unlike musical material—his "abruptness", in Wittgenstein's words—is a related obstacle that renders his music "unconnected" for many listeners. The fragmented introduction of the first movement of the third symphony begins to make sense once we attain a clear grasp of the typically Brucknerian so-called "redemptive" narrative of the symphony, especially of the fact—pointed out indirectly by Wittgenstein—that the first and the last movements must be regarded as logical sequels, indispensable and supplementary to each other (thus, indeed, as Wittgenstein said, the third movement does not lead into the fourth). The "redemption" lies simply in the success in securing tonic closure. Bruckner's imaginative unorthodoxy with regard to the key schemes of his sonata form and his formal strategic innovations enables him to postpone the definitive

63 Quoted in Michael Nedo, "Wittgenstein, die Musik und die Freundschaft" in Bruna Bocchini Camaiani and Anna Scattigno (eds.), *Anima e paura: Studi in onore di Michele Ranchetti* (Macerata: Quodlibet, 1998), p. 106. This letter, officially proclaimed to be unpublished and inaccessible, is in the possession of the Austrian National Library in Vienna.

64 See Leon Plantinga, *Romantic Music: A History of Musical Style in Nineteenth-Century Europe* (New York: W. W. Norton, 1984), pp. 435–440.

arrival on the tonic. Only in the coda, which remains outside the sonata space proper, can the triumphant tonic be reasserted and, in terms of the narrative, bring about redemption (hence the considerable importance of the coda in a Bruckner symphony).

The upshot is this: acknowledging that Bruckner's "abruptness" is an essential part of his musical language, hearing his music as a "connected whole" rests on an overview of the tonal narrative of the work with its uncontested, inevitable directionality. Here we come to a profound difference between Bruckner and Mahler, and to the reason for Wittgenstein's contention that "a symphony by Bruckner is infinitely closer to a symphony from the heroic period than is one by Mahler." As noted before, for some aspects of Mahler's music there are precedents in Bruckner; yet Mahler's approach to large-scale form was completely and radically new. Whereas a Bruckner symphony still exemplifies a closed system of musical relationship, Mahler introduced an innovative conception of musical form as a developing succession of individual episodes, held together by a complex network of interchangeable tonal relationships and by an elaborately developed system of motivic correspondences. This more open conception of form enabled Mahler to incorporate materials whose extreme contrasts would destroy the internal consistency of a more traditional context.

Georg Henrik von Wright recalled that Wittgenstein said that there was something initially incorrect in the architecture of Mahler's music.[65] Taken at face value—as when the term "architecture" is straightforwardly taken to denote large-scale form—Wittgenstein's observation may sound quite trite, and perhaps it is. Still, one is obliged to consider that the simile of "architecture" is not uncommon also in Wittgenstein's *Nachlass*. In particular, he refers to the architecture of mathematical systems, making the general point that a mathematical proposition can carry any weight, and is of any use, only insofar as there exists also a practice; otherwise it is no more than "a free-floating piece of mathematical scaffolding" (see MS 121, 41–42). From Wittgenstein's point of view, when compared to Bruckner's "way of telling the story," Mahler's way consists in precisely such a flawed architecture: if Bruckner is said to have composed in "main clauses," then Mahler must have been composing, oddly enough, *only* in "subordinate clauses."[66] Thus understood, Mahler's musical gestures veer away from their vertical axis; they become a kind of musical *Scheinarchitektur*.

---

[65] See note 59 above.

[66] It is noteworthy that Mahler's critics fiercely condemned precisely this character of Mahler's music, calling his works "gigantic symphonic potpourries." In a eulogizing essay on Mahler, Schoenberg explained: "The characteristic of the potpourri is the unpretentiousness of the formal connectives. The individual sections are simply juxtaposed, without always being connected and without their relationships (which may also be entirely absent) being more than mere accidents in the form." (Schoenberg, *Style and Idea*, p. 462)

Mahler's way of altering the meaning of tonality itself by means of his compositional procedures yielded music that may be fairly regarded as constructed. One may recall, in this context, Mahler's comment that to him writing a symphony means *constructing* a world with all the *technical* means at his disposal.[67] Such considerations illuminate Wittgenstein's assertion that Mahler is most unbearable precisely when he writes something that appears to be a perfectly grammatical musical phrase, for it is precisely in Mahler's deceptive simplicity—not in his embellished harmony—that the constructed nature of his music becomes painfully acute. To use Wittgenstein's own words, when writing in simple harmonic progressions, Mahler only *appears* to bear witness to Beethoven's or Bruckner's ancestral mother; in reality, since its rules of grammar are radically altered, Mahler's music bespeaks different things, involving concepts that are different, and ultimately, if it is a work of art at all, "it is one of a *totally* different sort." Thus Wittgenstein's point in reproaching Mahler—"you merely heard this from the others, this does not (*really*) belong to you"—begets its real philosophical thrust in a way that underscores the striking depth of Wittgenstein's ambivalence toward Mahler's musical *persona*: these musical gestures are merely *Scheinarchitektur*, not genuine avowals of the life of mankind—for how could they be?—and in this sense they are not authentic (*unecht*).[68]

We may conclude that the crux of Wittgenstein's hostility toward Mahler, of all other contemporary composers, was not atonality in itself, but rather the constructed nature of his musical language. As I suggested before, in a sense, atonality per se was simply uninteresting for Wittgenstein. It was not even a problem. The following diary entry from January 27, 1931 lends further support for these claims:

> The music of past times always corresponds to certain maxims of the good and the right of that time. We recognize Keller's principles in Brahms, etc, etc. Thus good music, which is being conceived today or has been conceived recently, that is to say modern, must seem absurd; for if it corresponds to any of the maxims *pronounced* today, then it must be rubbish. This sentence is not easy to understand but this is how things are: today no one is clever enough to formulate what is right [*das Rechte*], and *all* formulas and maxims that are pronounced

[67] See Natalie Bauer-Lechner, *Recollections of Gustav Mahler*, ed. P. Franklin (Cambridge: Cambridge University Press, 1980), p. 130.

[68] In a passage from 1938, Wittgenstein wrote: "Lying to oneself about oneself, lying to oneself about one's own inauthenticity [*Unechtheit*], must have a bad influence on one's style; for the result will be that one cannot discern within oneself between what is genuine and what is false. This may explain Mahler's style, and I am in the same danger" (quoted in Rush Rhees, "Postscript" in *Recollections of Wittgenstein*, p. 174). A further consideration of the deeply personal ethical tenor manifested in Wittgenstein's general attitude toward Mahler falls beyond the scope of the present essay.

> are nonsense [*Unsinn*]. The truth would sound *quite* paradoxical to everyone. And the composer who feels this within him must stand with this feeling in opposition to everything that is nowadays pronounced, and thus must seem by the present standards absurd, foolish. But not absurd in the *attractive* sense (for that is basically what the contemporary view corresponds to), but rather in the sense of *saying nothing* [*nichtssagend*]. Labor is an example of this, where he really created something important, as he did in some few pieces. (D, 38)

Wittgenstein presents three categories of contemporary music here: the good, the bad, and the meaningless. At least two of them—the first and the third—are genuinely intriguing from a philosophical perspective. According to Wittgenstein, bad modern music is conceived in accordance with prevailing contemporary principles, which are equally ill conceived. Most probably, Wittgenstein refers here to the predominant maxim of progress for which he had the deepest mistrust, as I noted before. Such was indeed the case with the emancipators of the dissonance in the name of progress during the first two decades of the twentieth century, and Wittgenstein clearly had no patience with their senseless musical gesticulation.[69] In his view, such music was plain rubbish.

Josef Labor exemplifies the intriguing category of the meaninglessly absurd. As I mentioned above, Labor, a protégé of the Wittgenstein family, was perhaps the only contemporary composer who won kudos from Ludwig Wittgenstein.[70] It is reasonable to say that even the best of Labor's music must have seemed absurd by the prevailing standard of progress. Indeed, against the background of the musical scene of *Fin-de-Siècle* Vienna, Labor's ultra-conservative, through-and-through tonal music gives the impression of having been composed in a time warp. Some of Wittgenstein's later remarks on Labor (dated approximately 6 months after the one just quoted above) corroborate this impression of inadequacy in Labor's music:

> Labor's seriousness is a very late seriousness. (MS 110, 231 – CV, 10)

> Labor, when writing good music, is absolutely unromantic. This is a very remarkable and significant characteristic. (MS 111, 2 – CV, 13)[71]

If we may recognize Keller's poetry in Brahms's themes, and if there is objective significance to the fact that these two men lived at the same time, as

[69] David Pinsent entered in his diary on October 4, 1912: "The second half of the concert began with two selections from Strauss's *Salome*: Wittgenstein refused to go in for them, and stayed outside till the Beethoven, which followed." Quoted in McGuinness, *Wittgenstein: A Life*, p. 124.

[70] Wittgenstein even attempted, albeit unsuccessfully, to promote a performance of Labor's string quintet in Cambridge. See ibid., p. 125.

[71] The fact that both remarks, written a week and a half apart, were entered in code signifies that Wittgenstein considered them sensitive.

Wittgenstein suggested (cf. LC, 32), then, quite conversely, we may experience in Labor's meaningless absurdity, in the fact that such music is seriously composed so very late, "a dissolution of the resemblances which unite [a culture's] ways of life," to use G. H. von Wright's words.[72] In other words, Labor's music lends an experience of an aspect of decline.

A further observation is in place here. It may seem as if Wittgenstein simply took sides in the great musical dispute that pervaded *Fin-de-Siècle* Vienna between Brahms's supporters and Wagner's enthusiasts. His clear rejection of progressive music seems to place him squarely among arch-conservatives such as composer Josef Labor and music critic Eduard Hanslick. Hanslick, like Labor, was closely associated with the Wittgenstein family; he was the most outspoken champion of Brahms's music in Vienna and the fiercest critic of Wagnerian innovations. Yet, as one would expect, Wittgenstein's position is ultimately much more complex and fine-shaded. His great admiration for Brahms's genius notwithstanding, Wittgenstein was still highly critical of some aspects of his music. His various comparisons between Brahms and Bruckner, for example, in which he points out that Brahms's music lacks orchestral color, suggest the convictions of a true *Wagnerianer* (see D, 44, 55–56). In fact, this kind of critique was fairly widespread among Brahms's detractors at that time. Still, Wittgenstein's most striking remark concerning Brahms was: "Music came to a full stop with Brahms; and even in Brahms I can begin to hear the sound of machinery."[73] Here, once again, Wittgenstein expresses a familiar train of thought held by others, ultimately traceable back to Heinrich Schenker, who felt that the great tradition of Austro-German music had come to an end with Brahms.[74] My point is this: by conceiving Brahms's music as a kind of zenith in the development of music, Wittgenstein sets the grounds for rejecting both the progressive approach and the conservative approach as viable options. Thus Wittgenstein's position actually transcends the Brahms-Wagner controversy. Labor's noble yet meaninglessly absurd rehash of classicism and Strauss's base, contrapuntal tinkering with harmony are both symptomatic of cultural decline.

This leaves us with the last alternative—good modern music—which, according to Wittgenstein, is actually no alternative at all. Incommensurability entailed by the concept of cultural decline renders the endeavor to create good modern music an absurd, albeit, according to Wittgenstein, an *attractive* absurd. One cannot, or at least one is not clever enough to formulate the right maxim or principle for our times—for what principle could be coherently pronounced amidst a dissolution of the resemblances which unite a culture's ways of life?—so, *ipso facto*, one cannot conceive of music that would corre-

[72] Georg Henrik von Wright, "Ludwig Wittgenstein in Relation to his Times," pp. 116–117.

[73] Quoted in Maurice O'C. Drury, "Conversations with Wittgenstein" in Rhees (ed.), *Recollections of Wittgenstein*, p. 112. This remark is dated back to 1930.

[74] See Heinrich Schenker, "Johannes Brahms," *Die Zukunft*, 19, 1897, pp. 261–265. I am indebted to John Daverio for this insight.

spond to the unpronounced. Thus, the precious little that Wittgenstein has to say about the category of good modern music is that this very notion is paradoxical.

Granting that contemporary music is a futile project *tout court*, what is left for a music of the future? Wittgenstein's tentative answer betrays, once again, a deeply Spenglerian vein:

> I should not be surprised if the music of the future would be monophonic [*einstimmig*]. Or is this just because I cannot clearly imagine several voices? In any case, I cannot imagine that the old *large* forms (string quartet, symphony, oratorio, etc) could play any role at all. If something like this comes, it will have to be—I believe—simple, *transparent*. In a certain sense, naked. Or will this apply only to a certain race, only to *one* kind of music (?) (D, 31)[75]

For Spengler, the future is always transcendent to the current epoch—"only youth has a future, and is future", he wrote[76]—and it is always marked by a return to the simplest, most basic expression of life.[77] A passage that Wittgenstein wrote in September of 1931 echoes Spengler's cyclic conception of cultural rejuvenation: "The works of great masters are suns, which rise and set around us. The time will come for every great work that is now in the descendent to rise again." (MS 111, 194 – CV, 15) We may, then, understand Wittgenstein's notion of the music of the future as the transcendent beginning of a new cultural epoch, hence the rejuvenation of music as a genuine avowal of the life of mankind. Thus Wittgenstein's position regarding the music of the future is consistent with his rejection of the aforementioned three categories of contemporary music, which are all immanent in the declining present epoch.

Wittgenstein envisions that a return to musical meaningfulness would take the form of monophonic music, or music in unison. Monophony, as distinguished from either polyphony or heterophony, simply means music for a single voice or part. Yet it is crucial to emphasize that the term monophony

[75] In order to understand this obscure, somewhat unfocused passage correctly, three points should be taken into consideration. First, Wittgenstein's remark concerning his inability to imagine several voices should be placed here in brackets. Later passages addressing the issue of contrapuntal music suggest that the problem concerning imagining several voices is not related to the present issue (cf. MS 163, 54 – CV, 40). Second, the terms "string quartet", "symphony" and "oratorio" do not denote musical *forms* in any technical sense (the notion of symphony as a musical form is ambiguous at best). Rather, they denote musical *formats* that are intrinsically related to a broad and highly complex cultural context (cf. LC, 8). Third, Wittgenstein's final, distinctly Spenglerian qualification should also be placed in brackets. This comment anticipates Wittgenstein's conceptual concession to cultural relativism, and as such it is irrelevant to our present concerns.

[76] Spengler, *The Decline of the West*, vol. 1, p. 152.

[77] See Spengler, *The Decline of the West*, vol. 2, p. 435.

is not synonymous with an unaccompanied melody. A melody specifically exemplifies musical movement that is set within internal musical boundaries: we hear that it begins, that it ends, and that is moves from its beginning toward its end. It is a closed system, as Wittgenstein acknowledged (RPP I §647). In tonal music, this has largely, albeit not exclusively, to do with harmony. But a monophony can be melodious without having a melody. An obvious example of such unbounded musical movement is a plainchant, or Gregorian chant, which is also the standard reference for monophony. The context of the passage quoted above strongly suggests that such a pre-tonal monophony is precisely what Wittgenstein had in mind. First and foremost, Wittgenstein's special interest in the problem of understanding *Kirchentonarten* (church modes or Gregorian modes) is evident in the *Nachlass* (see, e.g., PR §124; RPP I §639; PI §535), and in fact, Wittgenstein's first discussion of aspect perception in relation to music occurs in the *Philosophical Remarks* in reference to church modes.

Furthermore, by referring to something like a pre-tonal monophony as the music of the future, Wittgenstein echoes a broad intellectual concern regarding the putative origins of music that became widespread in central Europe from the turn of the twentieth century.[78] A brief historical excursion is required to substantiate this claim. Wittgenstein's Vienna was the intellectual cradle for the newly founded discipline of *Musikwissenschaft* (musicology): in 1870 Eduard Hanslick was the first to be offered a professorial chair in musicology. In 1898, musicologist Guido Adler was offered a professorial chair in Vienna. In his inaugural speech at the University of Vienna, Adler defined for the first time the agenda for musicological research in the German speaking universities for the years to come, establishing an archeology with which to reconstruct music history from its very first beginnings.[79] The interest in the origins of music also flourished in England, already during the second half of the nineteenth century: both Charles Darwin and Herbert Spencer offered evolutionary theories of music. These theories were received with great interest on the continent, and in 1911 psychologist and comparative musicologist Carl Stumpf published his book on the origins of music, in which he criticized Darwin and Spencer for failing to account for the specific features of music.[80]

In the same year, Wittgenstein arrived in Cambridge and engaged intensely in problems pertaining to the psychology of music under the supervision of Charles S. Myers.[81] Stumpf's work had a substantial presence in the milieu of the experimental psychologists in Cambridge: Myers himself used Stumpf's technical notion of "fusion" (*Tonverschmelzung*), sometimes without explicit

[78] See Alexander Rehding, "The Quest for the Origins of Music in Germany Circa 1900," *Journal of the American Musicological Society*, 53(2), 2000, pp. 345–385.

[79] See Guido Adler, "Musik und Musikwissenschaft," *Jahrbuch der Musikbibliothek Peters*, 5, 1898, p. 29.

[80] Carl Stumpf, *Die Anfänge der Musik* (Leipzig: J. A. Barth, 1911).

[81] See McGuinness, *Wittgenstein: A Life*, pp. 125–128.

reference[82]; the work of his colleague, C. W. Valentine, is replete with references to Stumpf's writings.[83] In Cambridge, Wittgenstein was exposed to Myers's own work on primitive music and the origins of music.[84] Myers clearly followed Stumpf in connecting the question concerning the origins of music with the question of how music evolved in the way it did. With these historical observations in place, a further important point is required to drive home my claim regarding Wittgenstein's reference to pre-tonal monophony as the music of the future. As Alexander Rehding observes, the search for the origins of music in the early twentieth century was not merely of archeological interest; it became instrumental in defining the tradition of tonal music as the subject matter of a science of music, not coincidentally, at a time when this tradition was increasingly perceived to be under threat from contemporary composition.[85] Against this backdrop, I suggest that Wittgenstein's advocating of monophony as the music of the future can readily be seen as a certain condensed version of this broad concern.

With this notion of pre-tonal monophony as the putative origin of music, we come full circle back to our use of language with its fine shades of behavior and meaning. One must acknowledge the fact that plainchant means primarily a vocal setting of a text: in liturgy, word and music are indissolubly connected. Instead of a melody in the modern sense, we have a series of inflections from a *reciting tone* that corresponds to the actual verbalization or vocalization of the text. According to Jeremy Yudkin:

> The music [in plainchant] is composed to words, which form grammatical units of sense, and the music reflects this sense. This does not mean that the music is "emotive" in the modern usage of the term, nor does it mean that the music indulges in "word painting" as in the Renaissance and Baroque eras (although instances of both of these practices can be cited). It means rather that in the clearest possible way the music is tied to the *structure* of the text, illuminating and clarifying the grammatical sense.[86]

To a large extent, the establishing of the reciting tone, the inflections and their range are vocal gestures, which are used like punctuation signs in a sentence.

---

82 See, e.g., Charles S. Myers, *In the Realm of Mind: Nine Chapters on the Applications and Implications of Psychology* (Cambridge: Cambridge University Press, 1937), p. 56.

83 See, e.g., C.W. Valentine, "The Appreciation of Musical Intervals," *The British Journal of Psychology* (1912).

84 See Charles S. Myers, "A Study of Rhythm in Primitive Music," *The British Journal of Psychology*, 1, 1905, pp. 397–406; and "The Beginnings of Music" in *Essays Presented to W. M. Ridgway* (Cambridge, 1913).

85 Rehding, "The Quest for the Origins of Music in Germany Circa 1900," pp. 371–380.

86 Jeremy Yudkin, *Music in Medieval Europe* (Englewood Cliffs, NJ: Prentice Hall, 1989), p. 43.

Plainchant epitomizes the "significant irregularity" that Wittgenstein points out as the hallmark of "phenomena akin to language in music" (MS 121, 26 – CV, 34).[87] The earliest chants must simply have been repetitions of a single pitch for every syllable of the text. Small inflections were added to mark the beginning of the whole reading and its end, the end of a sentence, or even a question form. Long segments were divided into smaller phrases by "musical commas"—endings that differ in pitch and formula from the ending of the whole sentence. Thus, a plainchant was originally an instrument of communication.[88]

Interestingly, a vivid impression of the sheer impact of the musical gestures of plainchant in relation to the spoken word at the very beginning of the Western musical tradition can be found in Saint Augustine's *Confessions*, a text that Wittgenstein knew all too well. Augustine had a tremendous impact on the acceptance of music into the church despite deep misgivings concerning the dangers in the musical obfuscation of language, which are given remarkable expression in book x of the *Confessions*. However, he was able to conclude that the power of music to convey the truth of the sacred texts separated it from the mindless pleasures, and his embrace of musical practices in his own services at Hippo—introducing into his worship several Ambrosian hymns and the Milanese antiphonal manner of singing—proved crucial to the development of Christian liturgy. One would imagine that Wittgenstein, being so repulsed by the post-Romantic excesses of his times, must have felt great sympathy for Augustine's advocacy of the simplest musical expression.

These considerations ultimately suggest that in Wittgenstein's vision of the music of the future we find a harbinger of both his later vision of musical expression, and his later general emphasis on language in use.[89] The history of music palpably teaches us that in plainchant we find the happiest marriage of music and spoken language. When Wittgenstein writes about the strongly

[87] It is noteworthy that historically the plainchant is a precursor of the recitative, which Wittgenstein brings as an example. Recitative differs from plainchant mainly in its precise rhythmic notation, its harmonic support, its wide melodic range and its affective treatment of the words. In this respect, plainchant makes a far better example of the significant irregularity exemplified by phenomena akin to language in music.

[88] This characteristic was brought out most clearly by the great dispute in the sixteenth century concerning the practice of polyphonic composition. One of the main claims against polyphony—advanced, for example, by Galilei in his *Dialogo della musica antica e della moderna* (1581)—was that its unequal rhythms, melodies and tempi, or its mingling of voices, impede communication, and that only uncluttered voice can communicate clearly. See John Neubauer, *The Emancipation of Music from Language: Departure from Mimesis in Eighteenth-Century Aesthetics* (New Haven, CT: Yale University Press, 1986), p. 26.

[89] For a discussion of Wittgenstein's particular stress on speech, see J.C. Nyíri, "Wittgenstein as a Philosopher of Secondary Orality," *Grazer Philosophische Studien*, 52, 1996/1997, pp. 45–58.

musical element in verbal language, he speaks of "a sigh, the intonation of voice in a question, in announcement, in longing; all the innumerable *gestures* made with the voice" (Z §161). It is in this flux of finely shaded intonation, Wittgenstein tells us, that we experience the meaning of words and make aesthetic judgments about them (cf. LC, 4). The music of the future is destined to be transparent precisely in the sense that sadness is transparent in a face; ideally, it is destined to be a physiognomy. Indeed such conception of "intransitive transparency" makes Wittgenstein's alternative metaphor, that of "nakedness," more apt.

Music for the Meaning-Blind

The history of twentieth-century music shows that it was Schoenberg, perhaps more decisively than any other composer of his time, who set sail toward a sonic landscape that became, at least for a while, the unmistakable music of the future. In the aftermath of the Second World War, Schoenberg's notion of the 12-tone system, rigorously emulated and applied by Schoenberg's most devoted disciple, Anton Webern, served as a catalyst for the young post-war generation of composers on both sides of the Atlantic Ocean for breaking completely with common compositional practice. This "music of the future" even has a name: total (or integral) serialism. The name refers to the fundamental conception behind this music, which was a consistent treatment of *all* musical elements—pitch, rhythm, dynamics, texture, and ultimately, form itself—according to strictly serial procedures, resulting in a complete departure from previous musical assumptions and traditional musical gestures. Consistent application of this idea brought about also—in what I am tempted to dub as an act of oedipal instinct—the ultimate abrogation of the Schoenbergian principle of the 12-tone row, which was ironically conceived as a relic of the "Old World" by prominent avant-garde composers of the second half of the twentieth century, such as Pierre Boulez, Karlheinz Stockhausen and John Cage.

Our discussion has so far suggested that given the Spenglerian *cum* Schenkerian forces found at play in the background of Wittgenstein's attitude toward contemporary music, we may expect nothing short of an insurmountable chasm between Wittgenstein the cultural pessimist, who admitted that he belongs together with Spengler "to the same group that is characteristic of these Times" (D, 28), and Schoenberg, the quintessential modernist, who avowed that he might be regarded conservative insofar as he conserves progress.[90] The precise nature of this chasm and its philosophical depth will now have to be made clear. Thus, the question before us is straightforward:

[90] Indeed, in an angry little essay from 1923 titled *Untergangs-Raunzer* or "decline-whiners" (an obvious allusion to Spengler's *Untergang des Abendlandes*), Schoenberg lashed out at "all these Spenglers, Schenkers, and so forth," who live the life of intellectual parasites, feeding on the works of art that they oppose. See Schoenberg, *Style and Idea*, pp. 203–204.

how far removed is Schoenberg's 12-tone music from Wittgenstein's vision of the music of the future? To approach this question, a still closer look at the genesis of Schoenberg's conception of the 12-tone system is required.

It is crucial to realize that the origins of atonality in Schoenberg's music are already deeply seated in his theoretical approach to tonal music. As Ethan Haimo points out, this can be seen in a variety of ways in Schoenberg's *Harmonielehre* (1911): he treats harmonic progression as not defining or establishing the tonic as the referential sonority; in his view, the diatonic collection does not define a key; and his explanations of chord formation blur the distinction between dissonance and consonance.[91] Tonality for Schoenberg was ultimately something of a contingency, causally explainable and susceptible to progress. Of course, such a conception of tonality betrays the concerns of a progressive composer, rather than those of a theorist speaking apparently of an aspect of the past. As Leon Botstein observes, this theoretical stance was at the heart of Schoenberg's confrontation with Heinrich Schenker:

> The crucial point of comparison between Schenker and the young Schoenberg was their shared conviction that music, although independent of words, operated by laws that were analogous to those of linguistic grammar ... The divergence between the two men rested on their assumptions about the possible future range of evolution for musical grammar, and not on the principle that music required the use of formal structures adequate to its autonomous character. For Schoenberg, musical grammar had both a teleology and an evolutionary history. For Schenker its nature was fixed.[92]

The quarrel between the two men extended far beyond technical matters. At stake were diametrically opposed views of the musical mind, and ultimately, I suggest, diametrically opposed attitudes toward language.

Consider, for instance, their dispute over the issue of non-chordal notes.[93] Traditionally conceived, non-chordal notes—passing notes, suspensions, auxiliary notes etc.—differ from chordal dissonances in that their resolution does not involve a change of harmony. A non-chordal dissonance is therefore incidental, for it does not impinge on the harmonic progression; it yields a momentary sonorous effect without harmonic consequence. As Carl Dahlhaus pointed out, both Schenker and Schoenberg rejected this traditional view, but for opposite reasons: while Schoenberg thought that the notion of an *incidental* dissonance is a misnomer, Schenker denied that a dissonance might be in any sense *essential* to the harmony. Schenker's position can readily be understood in the light of the précis of his theory, which I provided in the

[91] See Ethan Haimo, "Schoenberg and the Origins of Atonality," in Brand and Hailey (eds.), *Constructive Dissonance*, pp. 71–86.

[92] Botstein, "Music and the Critique of Culture," op. cit., p. 17.

[93] See Carl Dahlhaus, "Schoenberg and Schenker," *Proceedings of the Royal Musical Association*, 100, 1973/1974, pp. 209–215.

second section of the present essay ("Aspects of Decline"). If hearing music is to be understood as an exfoliation of a fundamental harmonic structure, then passing notes are the layers, so to speak, cast off in the process.[94] Hearing music amounts to hearing *through* the non-chordal notes that inhabit the various articulated musical layers (foreground, middleground or background), even those yielding the sonically harshest vertical combinations. In Schenker's words: "It is as if a vacuum existed between the dissonant passing note and the stationary cantus firmus note."[95] That is, the mere sonorous effect of the dissonance, its mere acoustics, has no musical meaning.

By contrast, Schoenberg's approach manifests a remarkable obsession with the "logic" of the musical surface. He dogmatically maintains that no musical occurrence can be without significance for the context of the harmonic progression, even those fleeting moments that are virtually imperceptible. Thus, any dissonant harmony resulting from a passing note is actually a chord and should be rendered vertically and independently as an essential phenomenon, to wit, as an emancipated dissonance. Simply put, according to Schoenberg, there is actually no such thing as a non-chordal note. Yet while the harmonic plausibility of the emancipated dissonance was something of an established fact for Schoenberg, it eventually led him to a painstaking—at times, arguably unsuccessful—pursuit of a justification of the harmonic function of the emancipated dissonance, and indeed, as I said before, to an obsession with what we could aptly call the "surface grammar" of music.[96]

A further consideration of Schoenberg's attitude toward language—crucial for our discussion—brings us back to Karl Kraus, closing, in effect, a line of reasoning which I began by entertaining the alleged "Kraus Connection" between Wittgenstein and Schoenberg in the first section of the present essay

[94] Schenker actually used the term *Schichten* (layers) as a technical term, denoting the long-range, mid-range and short-range melodic trajectories, divided under the headings background, middleground and foreground.

[95] Heinrich Schenker, *Die Meisterwerke in der Musik* (Munich, 1926), p. 25; quoted in Dahlhaus, "Schoenberg and Schenker," p. 210.

[96] It is noteworthy that Schoenberg's decisive leaning toward function in contradistinction to Schenker's leaning toward ornament in accounting for non-chordal notes aligns Schoenberg with Adolf Loos in a way that sheds a new light on the apparent asymmetry between the two, which I sketched in the first section of the present essay ("Leads and Impasses"). As I pointed out, Loos did not regard architecture as an art, whereas music, at least since the nineteenth century, was regarded as the ultimate art, the art to whose condition all other arts should aspire to rise. Yet, if we acknowledge Schoenberg's quasi-linguist emphasis on function in his approach to music, as we must, then we should also acknowledge that, within the framework of the Krausian dichotomy between the urn and the chamber pot, Schoenberg's art seems to fall peculiarly on the side of the chamber pot rather than on the side of the urn. Here, I believe, lies the real asymmetry that obtains between Loos and Schoenberg; hence we should concede that, while endorsing Schoenberg's middle-period atonal music, Kraus and Loos nonetheless might have overlooked the true nature of Schoenberg's project.

("Leads and Impasses"). It is evident that, on his part, Schoenberg misinterpreted the true nature of Kraus's thinking about language. In 1911, the same year he published his *Harmonielehre*, Schoenberg wrote:

> One may let oneself be carried by language, but it carries only the man who would be capable, if it did not exist, of inventing it himself. "Language, mother of thought," says Karl Kraus—as wrongly as if he had said the hen is there before the eggs. And as rightly. For that is how it is in the real work of art: everything gives the impression of having come first, because everything was born at the same moment. Feeling is already form, the idea is already the word.[97]

From Karl Kraus's perspective, there is something misleading already in binding literature and music together as Schoenberg does. According to Kraus, there is an important difference between verbal art and the other arts:

> Why do people treat literature so insolently? Because they know the language. They would take the same liberties with the other arts if singing to one another, smearing one another with paint, or throwing plaster at one another were means of communication. The unfortunate thing is that verbal art works with a material that the rabble handles every day. That is why literature is beyond help.[98]

Yet the point is that music is not beyond help in this sense. We can see that Schoenberg gave Kraus's acerbic dictum, "language is the mother of thought [*Gedanke*]", a Romantic reading as a license (for the genius artist) to meddle with language if language proves to be inert. This is a blatant misreading of Kraus insofar as it ultimately renders the actual means of expression subservient to the notion of an idea or a thought.[99] This is a crucial observation for our present concerns: Schoenberg's understanding of music as language precisely in this sense, in addition to his conviction that tonality has exhausted its natural resources, set the course, already in his middle period, toward the ultimate application of this misunderstanding of Kraus—the 12-tone system.

As I noted before, by 1923 Schoenberg felt that he had exhausted the so-called "free atonal" style with its expressionist pretense. He then mobilized his forces to regain control over his own composition processes. Reflecting on his motivation to construct his 12-tone system, Schoenberg wrote:

> [t]he desire for a conscious control over the new means and forms will arise in any artist's mind; and he will wish to know *consciously* the laws and rules which govern the forms which he has conceived "as in a dream". Strongly convincing as this dream may have been, the conviction that these new sounds obey the laws of nature and of our

[97] Schoenberg, *Style and Idea*, p. 369.

[98] Kraus, *Half Truths and One-and-a-Half-Truths*, p. 64.

[99] Schoenberg's apparent appeal to a sort of Romantic "big bang" theory of artistic genesis does not justify the conclusion he draws from Kraus.

> manner of thinking—the conviction that order, logic, comprehensibility and form cannot be present without obedience to such laws—forces the composer along the road of exploration. He must find, if not laws or rules, at least ways to justify the dissonant character of these harmonies and their successions.[100]

Schoenberg's obsession with "logic" took the form of a relentless quest—both in his theoretical thinking and in his compositional practice—for musical coherence; coherence that was lost when tonality was dissolved. "In music," he wrote, "there is no form without logic, there is no logic without unity."[101] Schoenberg used the term "coherence" to designate relationships that justify connections or meaningful interactions between the components of a sonic object. His attempt to emulate language is most explicit in his focus on finding and devising "musical connectives," akin to connectives in logic, that, so he believed, regulate the element of fluency in music and clarify the logic of its formal progression. He maintained that musical material should be both coherent and varied: "The preservation of features constantly secures logic, and upon the presence or absence of these connectives is based the greater or lesser degree of *fluency*."[102]

Now Schoenberg's 12-tone method was designed expressly to provide both coherence and variation in the musical material. At the heart of the system there is the 12-tone row, which is an "abstract" structure, a set of potential relationships without any motivic content that is "logically prior" to the actual composition. The row is embodied in the actual musical details of a given composition: it determines the succession of pitches used in a piece, although it does not determine their registers or their durations, nor prescribe the textural layout of the music or its form. Schoenberg conceived the 12-tone row as a pre-compositional fund for motivic possibilities, whereupon springs its sense of musical omnipresence. Thus according to Schoenberg:

> The weightiest assumption behind twelve-tone composition is this thesis: Whatever sounds together (harmonies, chords, the result of part-writing) plays its part in the expression and in presentation of the musical idea in just the same way as does all that sounds successively (motif, shape, phrase, sentence, melody, etc.) and it is equally subject to the law of comprehensibility.[103]

In Schoenberg's philosophy of composition, the notion of coherence is complemented by the notion of comprehensibility. "Composition with twelve tones has no other aim than comprehensibility,"[104] declared Schoenberg. Comprehensibility in general refers to conditions that allow the listener to grasp some-

[100] Schoenberg, *Style and Idea*, p. 218.
[101] Ibid., p. 244.
[102] Ibid., pp. 287–288.
[103] Ibid., p. 207.
[104] Ibid., p. 215.

thing as a whole, to bind impressions together into a form. As Carl Dahlhaus pointed out, the notion of comprehensibility, as Schoenberg uses it, is ambiguous: it refers to the emancipated dissonance per se, and at the same time it implies that the dissonance has a real function in the harmonic context.[105] Either way, according to Schoenberg's somewhat circular formulation, a musical content is comprehensible when it is surveyable and suitably articulated; that is, when its components share such coherence among one another and with the whole, as would in general be required for comprehensibility. In other words, coherence is a necessary condition for comprehensibility, which in turn ultimately amounts to the listener's ability to analyze quickly, to determine components and their coherence.

The contrived nature of 12-tone composition, in contradistinction to tonal composition, gives this notion of comprehensibility primary importance. In his third *Gedanke* manuscript (1925), Schoenberg points out that while compositions executed tonally proceed so as to bring every occurring tone into a direct or indirect relationship to the tonic, 12-tone composition presupposes knowledge of these relationships and does not render them as a problem still to be worked out. In this sense, 12-tone composition works with whole "complexes" akin to "a language that works with comprehensive concepts [*umfassenden Begriffen*], whose scope and meaning as generally known are presupposed."[106] Comprehensibility pertains to our ability to grasp and retain such fixed "concept-complexes," and to follow their implications and consequences.

Let us return now to our primary question: how far removed is Schoenberg's 12-tone music from Wittgenstein's vision of the music of the future? In a sense, by 1923 Schoenberg appeared to be heading back to a conservatively systematized conception of music. Yet, while his dodecaphonic works are thought out and worked out musically, they draw their motivic material from a contrived source: the 12-tone row.[107] Schoenberg was painfully aware that there was no escape from total chromaticism; for him, the genie of dissonance, once emancipated, could never be returned to the bottle again. Schoenberg argued that the 12-tone system is a necessary step in the evolution of Western music, and he designed it for the sole purpose of replacing the structural differentiations formerly furnished by tonality. Thus Schoenberg's late period music actually exemplifies a phantom U-turn to the old Western tradition of

[105] Dahlhaus, "Schoenberg and Schenker," op. cit.

[106] Schoenberg, *The Musical Idea*, op. cit., p. 416.

[107] Schoenberg never wanted his "method for composing with twelve tones which are related only with one another" to be freed of the conditions in which it had been conceived or of the ethical implications which it embodied. In 1923, 2 years after he had already began experimenting with the use of 12-tone rows in his music, Schoenberg gathered 20 of his students in order to stress upon them that " 'you use the row and compose as you had done it previously.' That means: 'Use the same kind of form and expression, the same themes, melodies, sounds, rhythms as you used before.' " See Schoenberg, *Style and Idea*, p. 213.

composing, a deliberate, conscious leap beyond what had been regarded as the "natural fountain" of musical language; yet one that Schoenberg firmly believed would "insure the supremacy of German music for the next hundred years."

Not surprisingly, Schoenberg's hubris drew a vehement response from Heinrich Schenker:

> The great proof against Schoenberg is the people; they have never gone along with him and never will. There are not two summits in an art. Schoenberg has already experienced the one, a second, like the one now being cultivated, cannot blossom. Schoenberg produces a homunculus in music; it is a machine. Machines are supposed to be substitute for human strength, a surrogate. Now there are of course surrogates, such as the one for traveling, the automobile, but *never can there be a surrogate for the soul.* Such a complicated operation is not intelligible for it. ... The product of Schoenberg's machine shall not be used.[108]

Schenker's riposte remarkably encapsulates the main themes that comprise what I maintain would be an adequate Wittgensteinian response to Schoenberg's dodecaphonic music: a sense of transgression, of soullessness and of contrivance. I have already discussed the first theme in detail. Schoenberg's 12-tone system is undoubtedly a fully-fledged instance of a systematic deviation from the rules of harmony. Furthermore, as pointed out before, the system was conceived not only to dislodge tonality, but also to downright take over its status as grammar. Schenker contended that "the great proof against Schoenberg is the people." Yet a much deeper insight is gained along Wittgensteinian lines: there is simply no reason for the rules of 12-tone composition to be what they are, given the kind of beings we are, the purposes we have, our shared discriminatory capacities, and certain general features of the world we inhabit. The kind of musical distinctions called for by dodecaphonic composition—for instance, identifying a certain passage as based on a certain transposition of the inverted retrograde form of the original 12-tone row used in the given piece—are not just very difficult to make; they are simply not important in our lives, certainly not in the sense that questions and answers, introductions and conclusions are.

There is no wonder, then, that the rules of 12-tone composition aim at nothing other than creating the conditions of comprehensibility. Schoenberg's striving for comprehensibility inevitably recalls Karl Kraus's repartee: "The most incomprehensible talk comes from people who have no other use for language than to make themselves understood."[109] A comparison between Schoenberg's standard of comprehensibility and Wittgenstein's standard of transparency or "nakedness" points at their crucial difference. According to

[108] Quoted in Snarrenberg, *Schenker's Interpretative Practice*, p. 89 (my emphasis).
[109] Kraus, *Half Truths and One-and-a-Half-Truths*, p. 65.

Wittgenstein, a musical gesture is not transparent by virtue of the correct applications of "rules of transparency"; rather, its transparency resides precisely in their absence, indeed in the vacuity of the very notion of such rules. Transparency in this sense is not an epistemic notion. A musical gesture is transparent because it is already given to us with a familiar physiognomy, already vertically related to our world of thoughts and feelings, whereupon there is no sense in which we can say that it needs to be *made* comprehensible.

Soullessness and contrivance go hand in hand. Schenker's allegation that Schoenberg's music toils at becoming a surrogate for the soul is quite remarkable. For Wittgenstein, nothing that is premised upon exactitude, calculation and mechanism can said to be soulful, since our recognition and description of soulful expression, musical or otherwise, is informed with, and constituted by evidential uncertainty, or "imponderable evidence." The imponderability of this kind of evidence is significantly reflected in the way we attempt to communicate our *Menschenkenntnis* and in the measure of the success of our justifications. Here the contrast between transparency and comprehensibility comes to a head. As we have seen, Schoenberg's view of music as language is rooted in what he perceived as a Krausian license to invent auxiliary means of expression in order to solve a particular problem—to wit, to regain control over unruly atonality.

The 12-tone system is an extraordinary attempt to derive, through a series of manipulations, a wealth of material, complex and varied, from an initial pitch collection that, in itself, is pre-compositional, hence musically inert and barren. In a banal sense, the fundamental elements of tonal composition—for instance, the particular pitch collection that we call the diatonic scale—are also "logically prior" to the composition. A tune like "Twinkle, twinkle, little star" has the particular effects of movement, rest and closure that it has because we hear the first and the last tones of the diatonic scale as the "same tone," and because we hear a certain hierarchic relationship obtaining between the other tones in the scale. In Wittgenstein's view, this phenomenology is embedded in, and makes any sense solely in terms of praxis (ultimately, our ways of life). Yet the point is that in Schoenberg's 12-tone system, "pre-compositional" means primarily "a-gestural"; and the latter notion, if it means anything at all, denotes something lifeless, soulless (cf. PI §§284–285). It is in this sense that the 12-tone row in itself is musically inert and barren[110]; hence at least some musical gestures found in 12-tone music are contrived by means of deliberate, rule-governed manipulation of this sort of pre-compositional material.[111] The result, to use Schoenberg's own telling analogy, is to be likened

[110] The fact that we hear the first and the last tones of the 12-tone row as being the same is beside the point. In doing so, we merely hear an interval of an octave; a tonal phenomenon that the 12-tone system professes to undermine ultimately.

[111] Musical gestures that pertain to dynamics, form, performance practice etc. are excluded here.

to a language comprised of concept-complexes whose meaning is semantically rigid like labels or name tags.[112]

Schoenberg was painfully aware of the constructed nature of his music, and he tried to counterbalance this impression by appealing to a view of (real) art as an organic whole:

> The inspiration, the vision, the whole, breaks down during its representation into details whose constructed realization reunites them into the whole. But this other constructed music which I have mentioned, and of which I have already seen examples, is different. It does not set out from the vision of a whole but builds upwards from below according to a preconceived plan or scheme but without a truly visualized idea of the whole, and it works up the basic material anxiously and without freedom. So whereas I proceed from a vision, working out the details and fitting them out for the purpose they will have to fulfill—and these details do not exist without that purpose—truly "constructed" music works material up into a systematically arrived-at, synthetically presented whole, which did not previously exist. In the former case it was the details that did not exist before; but in the latter, the whole.[113]

There is an obvious premonition in Schoenberg's characterization of the "other constructed music", of the kind of music making that was to take center stage in Europe under the banner of "total serialism" around the time of his death in 1951. Yet Schoenberg's attempt to rebut the charges concerning the constructed nature of his own music on grounds of the primacy of the musical idea over the construction of means for its expression betrays, once again, his misreading of Karl Kraus. Thus his defense remains ineffective from a Wittgensteinian point of view; for what is infuriating from the Wittgensteinian perspective is not so much the alleged genesis of this kind of compositional practice, as its pretense to inherit music. We can learn this by analogy from Wittgenstein's famous remark on Esperanto:

> The feeling of disgust we get if we utter an *invented* word with invented derivative syllables. The word is cold, lacking in associations, and yet it plays at being "language". A system of purely written signs would not disgust us so much. (MS 132, 69 – CV, 52)[114]

The striking analogy between Lazar Ludwik Zamenhof's vision of an international auxiliary language and Schoenberg's vision of the music of the future

[112] Obviously, here Schoenberg is exposed as being in the grip of the so-called Augustinian picture of language, which has been the elusive target of Wittgenstein's philosophical attack in his *Philosophical Investigations* (cf. PI §1).

[113] Schoenberg, *Style and Idea*, pp. 107–108.

[114] Note that Wittgenstein jotted down this comment on 26.9.1946, only a day after writing one of his most elaborate passages on musical understanding.

has not evaded scholarship.[115] Both projects arose as an attempt to solve a particular problem by rational means, laying down publicly defined rules for generating diversified means for expressing ideas. Both projects set themselves to overcome an initial alienation by communities steeped in tradition and well versed in the old ways of expressing ideas, and ultimately both met a similar fate: to be embraced only by a small, albeit passionate elite. Still, the most striking characteristic that Esperanto and dodecaphonic music share is the decisive shunning of all local or contingent effects of intercourse among the elements that comprise an utterance. As we have seen, in the case of 12-tone music, this took the form of a complete and irrevocable exorcising of the effects of tonality. In the case of Esperanto, this took the form of construing a vocabulary and a syntax that are exemplarily regular, efficient, consistent and accessible (to Europeans, at least).

As J. C. Nyíri observed, Wittgenstein's nausea had to do not so much with contrivance as with use.[116] What seemed to him despicable about Esperanto was the fact that this is an invented language—learnable by memorizing an economical vocabulary and a few syntactic rules—that one might want to use poetically. Rudolf Carnap, a passionate champion of language planning in general and of Esperanto in particular, recounts Wittgenstein's vehemently negative response when he learned of Carnap's interest in the problem of an international language like Esperanto. "A language which had not 'grown organically'," wrote Carnap, "seemed to him not only useless but despicable."[117] It is significant to note in this context that Carnap was particularly enthusiastic about the poetic promise that Esperanto held.[118] Carnap recalls a performance of Goethe's *Iphigenie* in Esperanto translation as one of the high points of an international Esperanto conference, which he attended. "It was a stirring and uplifting experience for me," he wrote, "to hear this drama, inspired by the ideal of one humanity, expressed in the new medium which made

---

115 See, e.g., Joseph P. Swaine, *Musical Languages* (New York: W. W. Norton, 1997), ch. 6.

116 J. C. Nyíri, "On Esperanto: Usage and Contrivance in Language," in Rudolf Haller and Johannes Brandl (eds.), *Wittgenstein – Towards a Re-Evaluation, Proceedings of the 14th International Wittgenstein-Symposium*, part II (Wien: Verlag Hölder-Pichler-Tempsky, 1990), pp. 303–310.

117 Rudolf Carnap, "Intellectual Autobiography," in Paul Arthur Schlipp (ed.), *The Philosophy of Rudolf Carnap, The Library of Living Philosophers*, vol. XI (La Salle: Open Court, 1963), p. 26.

118 From the very beginning, Esperantists were concerned about the aesthetic aspects and values of their language over and above one of its initially professed objectives, which was to serve as a vehicle for scientific communication. In the first publication in Esperanto in 1887, Zamenhof had already published three poems, and since then, the growing original literature in Esperanto has served as a device for the elaboration and testing of the aesthetic rules implicit in the structure and principles of the language. See Pierre Janton, *Esperanto: Language, Literature, and Community* (Albany: State University of New York Press, 1993), ch. 4.

it possible for thousands of spectators from many countries to understand it, and to become united in spirit."[119]

It was this vain attempt—vain for its artificiality—at a "vertical leap" from the language game of information to the language game of expression that infuriated Wittgenstein (cf. RPP I §170; §888). In the case of *our* language—a language that had "grown organically"—such a "vertical leap" to an (intransitively) expressive use of words is actually quite mundane, and significantly so. In fact, this is precisely Wittgenstein's point in suggesting that "understanding a sentence is much more akin to understanding a theme in music than one may think" (PI §527ff.). We may experience the meaning of words as irreplaceable, the thought in the sentence as "something that is expressed only by these words in these positions" (PI §531). Not only poetic language, but all language may be "musical" or "soulful" in this sense. Still, the seemingly unruly distinctions we make in experiencing the meaning of words, and the various ways in which we justify these distinctions are vertically complex in the sense exemplified by musical gesture (cf. PI §533; LC, 40). Inability to make such distinctions or to understand these kinds of justification is the mark of what Wittgenstein calls "meaning-blindness" (PI II, 175–176, 210; RPP I §§189, 202–206, 243–250, 342–344). The meaning-blind are locked out of that familiar physiognomy, which makes language something that we understand, not as a sign for something else, not transitively, but rather intransitively, like music, as "an avowal of the life of mankind." They are not attuned with the rest of us, not mutually voiced with respect to our fine-shaded use of language and behavior. For them, something has meaning only as part of an agreed symbolism used to convey information by depicting particular states of affairs. Such inability marks a total failure to become acculturated.

Thus, by conceiving language as music, Wittgenstein makes a fundamental point: words and phrases in language strike us as meaningful quite independently of their ability to convey information, and this feat, marking the success of acculturation, ultimately presupposes the entire range of our language games. We can invent a language, says Wittgenstein, in which "a b c d e" means "The weather is fine," and we could certainly use such an invented symbolism to communicate information about the weather. Yet the difference between such an invented language and natural, "organically grown" language is this:

> [i]n the one I can't move. It is as if one of my joints were in splints, and I were not yet familiar with the possible movements, so that I as it were keep on bumping into things. (RPP II §259; Z §6)

A natural language is fine-shaded, containing a myriad of possibilities that open up with each nuance of tone, each hint of a smile, and with all those "innumerable transitions which I can make and the other [who is not a native speaker of the language] can't" (RPP I §1078). According to Wittgenstein,

[119] Carnap, "Intellectual Autobiography," p. 69.

this is how understanding a sentence is comparable with understanding a piece of music. By contrast, an invented language is rigid, spasmodic, cold and lacking in associations. Its vocabulary is "a-gestural" in the sense that we have "no objection to replacing one word with another arbitrary one of our own invention" (PI §530). Thus, an invented language is "soulless": all that we have are signs that are translatable into action by means of rules. For Wittgenstein, such a language "does not get far as an impression, like that of a picture; nor are stories written in this language" (Z §145).

The analogy between Esperanto and Schoenberg's 12-tone system yields a conclusive answer to the question how far removed Schoenberg's post-1923 music is from Wittgenstein's vision of the music of the future. From Wittgenstein's perspective, Schoenberg's 12-tone music would be music for the meaning-blind, modeled on a conception of language as an artificial edifice, whose conditions of meaningfulness primarily consist in deriving a wealth of forms from musically barren sonic material by means of rules of coherence and comprehensibility; a kind of music, whose very essence shuns the familiar expanse of our *Menschenkenntnis*, where tonal music roams (cf. CV, 8–9). An actual performance of such music for the meaning-blind, enfolded by the gestural bravado of classically trained musicians, would be as despicable from Wittgenstein's point of view as a theatrical performance of Goethe's sublime poetry in Esperanto—it would be akin to an acquaintance with a surrogate for the soul.

Conclusion

I began my essay with Stanley Cavell's suggestion that Schoenberg's idea of the 12-tone row is a serviceable image of Wittgenstein's idea of grammar. The terrain is now carefully charted, and Cavell's direction appears unwarranted. We have seen that the only possible way to yoke Wittgenstein and Schoenberg (albeit indirectly) is through the respective impact of Karl Kraus's vision of language on both men. Yet this connection proved to be antithetical: Wittgenstein got Kraus's idea that "language is the mother of thought" exactly right, whereas Schoenberg got it exactly wrong. For Wittgenstein, thought presents itself only in our use of language, and understanding music is an avowal of the human life that shows itself in the grammar of our language. Music may be said to be transparent by letting itself be understood in this sense. Thus barren and inert, there is nothing in Schoenberg's row, the pre-compositional repository of musical thoughts, and in our presumed ability to comprehend these thoughts, that could compare to the power of grammar—as Cavell so aptly put it—to reveal pervasive yet unforeseen conditions of our existence. Wittgenstein's aversion toward modern music was shown to be rooted in his penetrating philosophical insight into musical meaning, not easily dismissible as a mere manifestation of a conservative musical taste.

A final passage from Karl Kraus would be appropriate for an epilogue to our discussion of these two incompatible visions of the music of the future:

> My pointer turns backwards; for me, what has been is never complete, and I stand otherwise in time. In whatever future I roam, and whatever I take hold of, it always turns into the past.[120]

In their mature work, both Wittgenstein and Schoenberg heralded a return to language, yet in different senses. Wittgenstein sought after transparency, Schoenberg after comprehensibility; for the one the very idea of a surrogate for the soul was an abomination, for the other—a fountain of youth. And so they roamed in different futures.[121]

[120] Karl Kraus, *Rückkehr in die Zeit* in *Werke*, vol. VII: *Worte in Versen* (Munich: Kösel, 1959), p. 236. English translation is taken from Alexander Goehr, "Schoenberg and Karl Kraus: The Idea Behind the Music," *Music Analysis*, 4, 1985, p. 71.

[121] I am grateful to Inbal Alexandron-Guter, Jaakko Hintikka, Enzo De Pellegrin and to the late John Daverio for helpful suggestions concerning some of the issues discussed in the present paper.

The Crash of the Philosophy of the *Tractatus*: The Testimony of Wittgenstein's Notebooks in October 1929

Jaakko Hintikka

One of the most crucial developments in Wittgenstein's philosophy was his rejection of phenomenological languages in October 1929. His notebooks from that period show how he came to take that momentous step. In this paper, I will present a translation of the crucial entries together with a commentary. It will turn out that Wittgenstein's brief notebook remarks are in reality connected with all sorts of different ideas in his overall philosophical thought and serve to illustrate them, even ideas that are not directly connected with his change of his main position vis-à-vis the contrast of physicalistic and phenomenological languages.

First, it is in order to clarify the very contrast between phenomenological and physical language. Contrary to what a large segment of the philosophical community seems to imagine, the terms "phenomenological" and "phenomenalistic" do not mean the same. An object is phenomenological if it is the object of immediate experience. But what is it that can be directly experienced? Many philosophers believe that only our own impressions are directly accessible to us in this sense. They cannot accept phenomenology without accepting phenomenalism. But the Cambridge philosophers who Wittgenstein was following did not think so. For a thinker like Moore, in any experience we can distinguish its object from the experience as an event in our consciousness. And this object is, well, objective; it is a part of reality itself and not only a part of our consciousness. For instance, for Moore and Russell sense-data were the objects of immediate perceptual experience, but they were at the same time denizens of the physical world.

Thus for Wittgenstein, too, the choice was not whether in a logically analyzed language we are speaking of members of the real world or merely of our own impressions. The question is whether the objects we are speaking of in language can all be directly experienced. (We may have to add here the qualification "the objects we are still speaking of when the language in question has been fully analyzed.") This sense of phenomenology as a position that relies only on what is directly given to me is in keeping with the usage of the terms "phenomenology" and "phenomenological" in the science

E. De Pellegrin (ed.), *Interactive Wittgenstein*, Synthese Library 349, DOI 10.1007/978-1-4020-9909-0_5, © Springer Science+Business Media B.V. 2011

and in the philosophy of science in the early years of the twentieth century. This usage was familiar to Wittgenstein, and it was the self-confessed precedent of Husserl's use of these terms.[1] In this usage, phenomenological objects were not contents of our consciousness, but observable entities in contrast to unobservable entities postulated in science.

Wittgenstein's phenomenology nevertheless has further features characteristic to him. The most important one is the idea that when an object is given to me, that is, when it is a phenomenological object, its logical form is also given to me. This logical form governs the way in which that object can and cannot be related to other objects. These possibilities of objects being combined with each other into a fact determines our logic. In other words, the set of experientially given ways in which the objects of our experience can be combined with each other is the basis of our logic *apud* Wittgenstein.

Thus for the early Wittgenstein the basis of logic is diametrically opposed to what it is to most philosophers. Typically, logic is thought of as being a study of our most general concepts. For Wittgenstein, it is based on the most particular things, the logical forms of phenomenological objects.

Wittgenstein retains this part of his early thought even after his 1929 change of position. It is phenomenology that still supplies the range of logical possibilities to him.

> Physics differs from phenomenology in that it is concerned to establish laws.
> Phenomenology only establishes the possibilities. Thus, phenomenology would be the grammar of the description of those facts on which physics builds its theories.[2]

Wittgenstein expresses his point by characterizing the phenomenological as that which is free from everything hypothetical.[3] What this amounts to is the same direct givenness as I have attributed to him. For if an object is not given to me in direct experience, much of what I have to say about it is hypothetical in the sense that I could be wrong. I may be wrong in identifying the object in question and I may be wrong as to what it is like. I may be wrong in thinking that it exists. In the case of directly given objects, it does not make sense to doubt any such things.

In the *Tractatus* phenomenological objects are what the names of Wittgenstein's regimented language stand for. By choosing as the representatives of different phenomenological objects names which have the same logical form as their objects, we can have a language in which the totality of possible

[1] See *Husserliana: Edmund Husserl – Gesammelte Werke*, vol. 9, ed. Walter Biemel (The Hague: Martinus Nijhoff, 1968), pp. 302–303.

[2] Ludwig Wittgenstein, *Philosophical Remarks*, ed. Rush Rhees (Oxford: Basil Blackwell, 1975), p. 51. Page references to the *Philosophical Remarks* will be to this edition.

[3] See *Philosophical Remarks*, pp. 97–101.

combinations of names mirrors the totality of possible configurations of their objects in the world.

This also helps to explain the relation of Wittgenstein's tacit phenomenology in the *Tractatus* to other philosophers. In the phenomenological language of the *Tractatus* names stood for what were essentially objects of acquaintance à la Russell. The probable reason why Wittgenstein did not call them objects of acquaintance is that he wanted to highlight the differences between Russell and himself, especially the crucial fact that for him logical forms were not freestanding objects, as they were for Russell. The reason Wittgenstein did not in the *Tractatus* call his simple objects phenomenological is that he did not want to be associated with Ernst Mach's phenomenology. In spite of not inconsiderable similarities in their overt philosophical views, Wittgenstein considered Mach a superficial thinker.[4]

The other side of the conceptual coin is that by a "physicalistic" language I do not mean, and Wittgenstein would not have meant, a language of physics. What Wittgenstein has in mind is an everyday language in which we speak of such familiar friends as ordinary physical objects, persons and so on, not necessarily a language of unobservable subatomic particles or fields of force. The crucial point is merely that not all such commonplace objects are given to me in my immediate experience. Already in Russell, familiar persons like Bismarck or Caesar are not all objects of acquaintance for this reason. All this amounts to little more than saying that Wittgenstein's distinction between phenomenological and physicalistic languages was—as he makes clear himself—equivalent with other philosophers' contrast between "the primary system" and "the secondary system."[5]

After these preliminary explanations, we can address Wittgenstein's notebooks from the years 1929–1930.[6] Wittgenstein was in the fall of 1929 struggling with the problem of truth, meaning and of the objects that represent

[4] See my "Ernst Mach at the Crossroads of Twentieth-Century Philosophy" in *Future Pasts: The Analytic Tradition in Twentieth-Century Philosophy*, edited by Juliet Floyd and Sanford Shieh (Oxford and New York: Oxford University Press, 2001), pp. 81–101. See also Brian McGuinness, *Wittgenstein: A Life* (Berkeley and Los Angeles: The University of California Press, 1988), pp. 38–40.

[5] See *Philosophical Remarks*, pp. 58, 84.

[6] Microfilm copies of many of Wittgenstein's known notebooks from 1929 until 1932 have been publicly accessible and available for purchase since 1967 through the Cornell University library system. Parts of Wittgenstein's *Nachlass* were later published in the *Wiener Ausgabe* edited by Michael Nedo. A joint project of Oxford University Press and The Wittgenstein Archives in Bergen recently resulted in the publication of a series of CD-ROMs (*Wittgenstein's* Nachlass*: The Bergen Electronic Edition* (Oxford and New York: Oxford University Press, 2000)), which provide scholars with both a highly readable rendering of the German text of a large portion of Wittgenstein's *Nachlass* as well as with image files of the individually scanned manuscript and typescript pages. References to writings from Wittgenstein's *Nachlass* will be to this most recent edition.

and are represented. He is operating within a phenomenological framework, with phenomenological objects as the targets of linguistic representation. He was beginning to have doubts, however.

> It is as if the phenomenological language led me into a bewitched swamp where everything tangible disappears. (MS 105, p. 116, undated, but probably summer 1929)

But he has not given up.

> And yet there can be a phenomenological language. (MS 107, p. 3, undated, but probably summer 1929)

One reason Wittgenstein has for accepting the idea of a phenomenological language is that otherwise linguistic representation would be incomprehensible for him. This is based on the idea that language and reality must be immediately comparable.

> One cannot compare a picture with reality unless one can lay it down on it like a measuring rod.
>
> One must be able to put the sentence on [the top of] reality.
> [deleted alternative: be able to bring the sentence to coincide with reality.]
> (MS 107, p. 152)

But it is clear that the envisaged picture must itself be real, e.g. of the nature of an image, in order to be compared with reality. Hence we are no longer dealing with pictures in the usual sense. Hence, in a comparison between language and the world

> The envisaged reality replaces the picture. (MS 107, p. 152)

For instance

> If I am to ascertain whether two points are located at a certain distance from each other, I must grasp that distance with my eyes.
>
> I must after all *compare* the reality with the sentence. (MS 107, pp. 152–153)

And that presupposes a representation which precedes the comparison. In other words, I must form an envisaged reality which functions in the same way as a picture, in particular, which is eventually compared with reality.

When a proposition is not yet so to speak on the top of a fact, comparability presupposes that we can search for the relevant fact. We have to look around for the purpose of comparing the envisaged reality with the actual reality. This search is guided by the proposition (picture) or, rather, the envisaged reality which is to be compared with facts.

> One cannot search falsely, one *cannot* search for a visual impression by means of the sense of taste. (MS 107, p. 152, undated, probably October 4–5, 1929.)

Thus Wittgenstein's requirement of comparability leads him to inquire into the conditions on which one can be said to search for something. He is thereby relying on an idea which is enormously important but which he never seems to have developed, not even dragged out to the open. It is a connection between existing (being one of our *objects*) and being capable of being *found*, that is, successfully *searched for*; in other words a connection between being an object and being capable of being looked for and found. In terms that Wittgenstein might have used himself later, this could be expressed by saying that seeking and finding constitutes the language-games which are the logical home of the notions of existence and of (an existing) object. But in 1929 Wittgenstein had not yet developed his notion of language-game. And later, alas, after he had done so, he did not go back and relate the language-games of seeking and finding to the notions of existence and objecthood—nor as much as to conceive the activity of searching as a kind of language-game. Anyone who is familiar with the subsequently developed game-theoretical semantics is likely to shudder at the opportunities Wittgenstein missed here.

Why did he miss them? If a conjecture is allowed, I suspect that Wittgenstein was thinking of existence in terms of the existential quantifier and thinking of this quantifier as "ranging over" a set of values, in spite of his realization that the construal of existential sentences as disjunctions (and universal ones as conjunctions) was his "biggest mistake" in the *Tractatus*.[7] Wittgenstein might also have been preoccupied too much with the different methods—and different kinds of methods—of searching to accept the generic activity of searching as unequivocal enough to constitute a single language-game that could lend a meaning to quantifiers in general. This would have been in keeping with his constructivistic tendencies.

To some extent, the job of the language-games of seeking and finding was done in Wittgenstein's thought by the notion of space. The connection is obvious: seeking and finding can be performed only in some search-space. Visual space and color space are conditions for this role. The importance of the idea of space for Wittgenstein in 1929 is also shown that in the language envisaged in his transitional paper "Some Remarks on Logical Form" was one in which atomic sentences contained numerical parameters expressing coordinates in some space or other.[8]

---

[7] See Georg Henrik von Wright, *Wittgenstein* (Oxford: Basil Blackwell, 1982). p. 151.

[8] See the 1929 paper "Some Remarks on Logical Form", first published in the *Proceedings of the Aristotelian Society*, Suppl. vol. 9 (1929), pp. 162–171. Reprinted in *Philosophical Occasions. 1912–1951*, ed. James Klagge and Alfred Nordmann (Indianapolis: Hackett Publishing Company, 1993), pp. 29–35.

To return from such virtual history to the actual one, in his 1929 notebook Wittgenstein concentrates on the conditions of finding a looked-for object rather than on the conditions of search. More specifically, he focuses on the conditions of recognizing the object of search. According to him, this presupposes immediate comparability. In order to search for something I must have a way to envisage what I am looking for. Moreover, the eventual comparison between language and the searched for reality must be immediate.

Wittgenstein is thus suggesting that imagined and other envisaged scenes can and must represent reality and that ordinary pictures and sentences must be accompanied by such representations. This is presumably because such visual or other experienced images are the only objects that can be immediately confronted with reality. This is in turn due to the fact that the reality we speak of in our language is phenomenological, that is, immediately accessible to our experience. As Wittgenstein put it in his lectures, "the world we live in" is the world of "sense-data." Hence the "pictures" that our sentences are must also be at bottom phenomenological in order to be comparable to experienced reality.

> "Blue and white are next to each other", that is obviously a sentence, but obviously also a picture. (MS 107, p. 153)

In other words, pictures and sentence are on a par. But not in all respects:

> The sentence is not simply a picture, but a portrait. (MS 107, p. 155, October 7, 1929)

In other words, if a sentence is to express a specific meaning, it must speak of certain particular objects excluding their potential or actual "lookalikes." Wittgenstein's meaning is illustrated by his further comments on the same subject.

> Is the fact that there can be pictures [variant: that one can make pictures] that are not portraits connected with the fact that the world is a temporal one?

By pictures that are not portraits Wittgenstein means pictures that represent objects of a certain kind, not certain specific objects. Elsewhere he calls them genre pictures. Wittgenstein's idea apparently is that even though two exactly similar objects can be distinguished from each other when they are presented to me simultaneously by their merely being different, they cannot be told apart when presented to me at different times. Indeed, in *Philosophical Investigations* I, §253, Wittgenstein acknowledges that we are in a position to speak of two precisely similar objects,

> ... for example, to say "This chair is not the one you saw yesterday, but it is precisely similar to it."[9]

[9] Ludwig Wittgenstein, *Philosophical Investigations. The German text, with a revised English translation*, third edition, trans. G.E.M. Anscombe (Oxford: Black-

And of course two precisely similar objects can (and perhaps must) be represented by the pictures that look similar. In brief, after all, different objects can be indistinguishable when we meet them at different times, because they are associated with the precisely similar image or other representation. It is thus significant that Wittgenstein is considering two different moments of time here.

Wittgenstein's problem of the possibility of genre pictures is interesting here mainly because what it shows about his thinking in 1929. He is obviously thinking that any language has to be based on a correlation of particular names and specific objects. Where there are no specific objects represented by each of the ingredients of a proposition (picture), some further explanations are needed, as in the case of genre pictures. This illustrates indirectly the importance for Wittgenstein of his quest of the objects we are speaking of in statements of direct experience. All this is based on the idea that a judgment of similarity is one possible result of a comparison.

> My main idea is that one *compares* a sentence with reality. (MS 107, p. 155)

But how is that possible? Wittgenstein runs into a major difficulty.

> I cannot work properly, or at all, the philosophical region of my brain is still in the dark. And it is not until the light is again lit there that the work can again progress. (MS 107, pp. 155–156, October 8, 1929)

And what is the difficulty? Wittgenstein expresses it by means of an example, asking:

> What is meant by explaining the emphasis we put on something?
>
> Can one say: The emphasis expresses something that only it can express and what cannot be expressed without it? (MS 107, p. 156, October 8, 1929)
>
> [In English] Emphasis can only be replaced by emphasis, not by what is emphasized. (MS 107, p. 156, October 9, 1929)

What Wittgenstein is discussing here is a special case of a more general problem: Emphasis as a phenomenological object is not expressed by any feature of the "picture" that a sentence is. As a consequence, emphasis and what is emphasized are not commensurable. They cannot be compared directly with each other. Hence we have here a *prima facie* counter-example to the comparability that Wittgenstein has just demanded. Hence Wittgenstein concludes:

---

well Publishing Ltd., 2001), §253. Anscombe mistranslates "genau gleich" as "exactly the same", thus missing the crucial contrast between "derselbe" and "genau gleicher". References to the *Philosophical Investigations* will be to this edition.

> The problem of truth eludes me.
> I am aware that the most wonderful problems are close to me. But I don't see them and cannot grasp them. (MS 107, p. 156, October 9, 1929)

Yet Wittgenstein resumes his examination of how language and reality are connected by search processes. One crucial point he returns to in 1930 is the role of "pictures" in guiding our search for certain objects.

> You have seen a particular blue—say, sky-blue—and presently I show different patterns of blue to you. Then you say, "No, no, it was not this one, nor that, nor that.—Now, that is the one!" Is it as if you had various push-buttons in your head and I was trying them out until I pushed a particular one and then the bell rang? Does the recognition of a colour come about in the same way? Does a bell ring in me, as it were, does something click when I see the right colour? No! Rather, not only do I know that a particular blue is not the right one but I also know in which direction I have to alter the colour in order to reach the right one. This means, I *know a way of looking for this colour*. If you have to mix that colour I can give you hints by saying, "more white, still more white, no, that is too much, some blue, and so forth." That is, this colour presupposes the whole *colour-system*. Recognizing a colour is not simply a matter of comparison, although in some respects it is similar to a comparison.[10]

Thus speaking meaningfully of any object presupposes a kind of logical space (potential space of search) in which it is located. But where do I find such a space in the case of objects of private sensations and other objects of immediate experience?

> One cannot deny that one has a stomach-ache without immediately envisaging the[ir] possibility. But what does it mean here "envisaging the possibility"? A primitive representation would be that the stomach M were given to me in a certain context and the pains [the ache] S in another context and now I would come to see that M and S are not connected with each other.
>
> But of course this is not the case.
>
> I can very well also say: I see a red spot without seeing anything that is red. On the contrary, the possibility of red lies in the seeing of any color, hence in seeing as such. (MS 107, pp. 157–158, October 10, 1929)

[10] *Ludwig Wittgenstein and the Vienna Circle, Conversations Recorded by Friedrich Waismann*, ed. Brian McGuinness (Oxford: Basil Blackwell, 1979), pp. 87–88, dated January 5, 1930.

Comments: Thus color space is for Wittgenstein an example of the kind of realm of possible interrelations of objects determined by their logical forms that is given to me when I directly experience objects. Likewise, the possibility of meaningful negation entails that the objects of experience are situated in a space of alternatives. This is connected with Wittgenstein's earlier comment on searching. The space is what allows us to speak of a search there. I can speak meaningfully of not finding something only if I can also meaningfully speak of finding it. The possibility of searching was seen to be grounded on the phenomenology, i.e., the logic, of different dimensions of reality. The different elements of the search-space must be given to one in present or past experience. In fact, this is confirmed by Wittgenstein's next statement.

> The only essential thing is that I envisage a space in which the stomach is located and the one where the ache is located. But how can the stomach-ache space be present when I do not have stomach pains?

Elsewhere Wittgenstein expresses the same point as follows:

> If I say "I have not got stomach-ache," then this presupposes the possibility of a state of stomach-ache. My present state and the state of stomach-ache are in the same logical space as it were. (Just as when I say "I have no money." This statement presupposes the possibility that I do have money. It indicates the zero point of money-space.) The negative proposition presupposes the positive one and *vice versa*.[11]

One problem here is that the elements of the stomach-ache space must be given in experience, including memory. But I cannot have even memories of all possible pains (aches), for I have not experienced all of them. An even more general problem is: What can be meant by "space" here? What are the elements of different spaces? Wittgenstein continues the thoughts expressed in the preceding statement and takes an example where we do seem to be able to speak of space, viz. color space.

> In the same way as it makes sense to say that the color R is at a certain location P when I have "in front of me" the visual space in conjunction with the color space. But these two spaces are not on a par. For I can *carry out a search* in the visual space but not in the color space. I can look for a white spot on my suit but not look for the location on my suit in the color spectrum.
>
> But is that an essential difference? Isn't it also possible to look for something in the color spectrum. Let's assume that I have apparatus to create the colors of the rainbow on a strip one after another. Cannot I then look for the location where a certain [shade of] orange will

[11] *Ludwig Wittgenstein and the Vienna Circle*, p. 67.

appear?

> When a continuous spectrum is given to me and [likewise] the sentence "red is [to be found] at a certain place S" can't I then equally well look for red on the spectrum in order to test this sentence and see whether it is located at the [a certain] place S as I can look for the place S and see whether red is located there? (MS 107, p. 158, October 10, 1929)

Thus colors, and presumably by the parity of the two phenomenological cases stomach-aches, too, seem to be objects. But are they really? Wittgenstein is unable to answer his own questions. He realizes that he is not asking the right ones.

> Today I feel a special lack of problems around me, a sure sign that the most important and most difficult problems are *facing me.* (MS 107, pp. 158–159, October 10, 1929)

In the end, Wittgenstein concludes that in immediate experience there is no "space" of constant possibilities present.

> The immediately given is in a state of constant flux. (It has in fact the form of a river.) (MS 107, p. 159, October 11, 1929)

Hence there is no way of searching anything in the immediately given. Hence there cannot be any comparisons between pictures and reality in the immediately given, ergo, no representations. Wittgenstein has reached a dead end. What makes things worse for Wittgenstein is that, according to his view,

> [o]ur propositions are only verified by the present. (MS 107, p. 222)[12]

He expresses his resignation in his own characteristic words:

> It is clear that if one wants to say the last word here one must come directly to the limit of the language that expresses it. (MS 107, p. 159)

Hence Wittgenstein must proceed in a new direction.

> The worst philosophical mistakes come always about when one wants to apply our usual—physical—language in the field of the immediately given. (MS 107, p. 160)

This is the first time Wittgenstein asserts the conceptual primacy of physicalistic language. Wittgenstein goes on to spell out his new insight.

> If one asks e.g. "does the box still exist when I do not look at it?" then the only right answer would be "of course if nobody has carried it away or destroyed it." Of course a philosopher would not be satisfied by this

[12] See also *Philosophical Remarks*, p. 81.

answer, but it would quite correctly lead his questioning ad absurdum.

> All our ways of speaking are borrowed from the normal physical language and are not to be used in epistemology or phenomenology without putting the subject to a wrong light.
>
> Even the bare locution "I perceive x" is borrowed from the physical[istic] way of expressing oneself, and [hence] x should be here a physical object—e.g. a body. It is already mistaken to use this locution in phenomenology where x must then be a [sense] datum. For now the locution cannot either have the same sense as [in the physical locution] above. (MS 107, p. 160)[13]

A few paragraphs later Wittgenstein writes:

> These are the dangerous shifts of meaning "I hear music", "I hear the piano", "I hear him playing the piano." (MS 107, p. 161)

This is indeed a momentous change in Wittgenstein's position. What it means is that the objects we are speaking about, what our words and other basic symbols refer to, are not given to me in experience even when we are describing our immediate experiences. They are denizens of the everyday physical world. They need not be given to me in immediate experience, and their identities need not be known to me. (This is the reason why the treatment of identity in the *Tractatus* does not any longer interest Wittgenstein.) Hence, for one thing, there will be singular statements about our linguistically and semantically basic objects that are merely hypothetical. For the absence of everything hypothetical was seen to be the characteristic mark of phenomenological languages *apud* Wittgenstein's new sense of hypothesis.

> An hypothesis is not a statement, but a law for constructing statements.[14]

Hence Wittgenstein's new notion of hypothesis is an immediate consequence of his rejection of phenomenological languages. Later, Wittgenstein accuses Carnap of stealing this notion of hypothesis from him.[15]

Another consequence of what happened on October 11 is that it becomes difficult for Wittgenstein to maintain any direct comparability between language and the world, as he would like to do. This creates a persistent tension in Wittgenstein's thinking. He is reduced to making the comparability a matter of relations of entire language systems and reality.[16]

---

13 See also *Philosophical Remarks*, p. 88.

14 *Ludwig Wittgenstein and the Vienna Circle*, p. 99.

15 See my "Ludwig's Apple Tree: On the Philosophical Relations between Wittgenstein and the Vienna Circle" in *Ludwig Wittgenstein: Half-Truths and One-And-A-Half-Truths. (Jaakko Hintikka Selected Papers vol. 1)*, edited by Jaakko Hintikka (Dordrecht: Kluwer Academic, 1996), pp. 125–144.

16 As can be seen, e.g., in *Ludwig Wittgenstein and the Vienna Circle*, pp. 213–215.

The switch from phenomenological to physicalistic basic languages thus poses problems for Wittgenstein. He no longer complains about a dearth of philosophical problems. Indeed, virtually every extension of a physicalistic term to the phenomenological realm in philosophy will now require a re-examination, for such extensions usually cannot be taken in a literal sense. A case in point, taken up by Wittgenstein later, is the talk about ambiguous figures and about aspect seeing.[17]

Some of the new problems are direct consequences of the problems that led Wittgenstein to change his basic languages. One problem that hit him immediately after proposing to make physical languages primary is where to find the objects we mean when we speak of geometrical objects. There are apparently such phenomenological entities as perfect circles and perfectly straight lines. However, no actual physical circles or lines are perfect. If the ontology that our language relies on is restricted to physical objects, what are we talking about when we speak of perfect geometrical figures? Wittgenstein realizes that in the two cases, viz. seeing a perfect geometrical object and seeing a physical object we are using different senses of seeing.

> If I cannot see a precise circle than I cannot in the same sense see any approximate circle.—Then the Euclidean circle—as well as an approximation to the Euclidean circle—is not in this sense an object of my perception at all, but, say, only a different logical construction which could be obtained from the objects of a quite different space from the space of immediate vision. (MS 107, pp. 161–162, October 11, 1929)[18]

Wittgenstein struggles in the next few days with the problem of unsharp representation—and of the representation of unsharpness. Finally he solves these problems well enough to be ready to draw a definitive conclusion:

> The assumption that a phenomenological language were possible and that only it would express what we must [Wittgenstein's alternative: want to] express in philosophy is—I believe—absurd. We must get along with our usual language and merely to understand it correctly, that is, we must not be seduced from it to speaking nonsense. (MS 107, p. 176, October 22, 1929)[19]

---

[17] I have examined this topic (jointly with Merrill B. Hintikka) in our 1985 paper "Ludwig Looks at the Necker Cube: The Problem of 'Seeing As' as a Clue to Wittgenstein's Philosophy", reprinted in *Ludwig Wittgenstein: Half-Truths and One-And-A-Half-Truths*, pp. 179–190.

[18] See also *Philosophical Remarks*, p. 265.

[19] This conclusion can be compared with a similar announcements on p. 88 of *Philosophical Remarks* and on p. 45 of *Ludwig Wittgenstein and the Vienna Circle* (dated December 22, 1929).

Hence one of the mainstays of Wittgenstein's philosophy in the *Tractatus*, his reliance on phenomenological languages, crashed the same week as the Wall Street—indeed, 2 days before the Black Thursday.

The situation of language priorities entailed profound changes in Wittgenstein's methodology and in a sense in his ontology, at least in the relation of the ontology to language. For one thing, an important consequence of Wittgenstein's change of mind is there is no longer one privileged mode of linguistic representation, viz. by way of representing the given. In the *Tractatus*, in Waismann's words,

> Elementary propositions describe the content of our experiences.[20]

> Phenomena (experiences) are what elementary propositions describe.[21]

Wittgenstein discusses hypotheses in MS 109, pp. 16–23 (dated August 15–22, 1930). This discussion confirms what has been said here. What is expressible in language are in the first place "hypotheses", not immediate experiences.

> Experiences, that is, primary events, are *compatible* with the hypothesis. [In English:] (The hypothesis *accounts* for them.)
> One could say something like: the hypothesis *explains* them.
>
> Isn't it now the case *that* what a hypothesis explains is again expressible only by a hypothesis. That is, naturally, are there any primary propositions that are conclusively verifiable and not only aspects of a hypothesis? (This is like asking: "Are there surfaces that are not surfaces of bodies?") (MS 109, pp. 19–20, August 18, 1930)[22]

We can now see that Wittgenstein's change of his language paradigm was just that: a change in what he took our actual language to be like. His view of reality remained phenomenological. As we saw from the quote from MS 109, pp. 19–20, the basic events were for him experiences. By the same token, the basic reality consisted of the objects immediately given to me.

> The world we live in is the world of sense data; [...][23]

Instead of "sense data" Wittgenstein could, and perhaps should, have said more generally "phenomenological objects."

What is more, these phenomenological objects still determine the logical structure of the world for Wittgenstein. As he put it, for him *phenomenology is grammar* or, as we would prefer to say phenomenology is logic. And this

---

[20] *Ludwig Wittgenstein and the Vienna Circle*, p. 254.

[21] *Ludwig Wittgenstein and the Vienna Circle*, p. 249.

[22] For the last paragraph, see also *Philosophical Remarks*, p. 221.

[23] *Wittgenstein's Lectures: Cambridge 1930–32. From the Notes of John King and Desmond Lee*, ed. Desmond Lee (Oxford: Basil Blackwell, 1980), p. 82.

statement is intended to be convertible: logic is at bottom a phenomenological matter.

But that logic is not and cannot be the logic of our language. As Wittgenstein continues the last quote,

> [...] but the world we talk about is the world of physical objects.[24]

Thus the true logical structure of the world cannot be codified in the logical syntax of any language, natural or ideal. It cannot be obtained by analyzing any language, ideal or natural. (In this sense, the later Wittgenstein was not any more a "philosopher of ordinary language" than the early one.) The true representation of reality, that is, a representation of immediate experience can be studied only indirectly by comparing different methods of representation with each other to see how they do their job.

Thus the change in Wittgenstein's entire philosophical methodology is due to his switch from phenomenological to physicalistic languages. It is not due to a rejection of the "picture theory" or to any realization that language can be used in many different ways. In a sense, it does not even mean giving up the aim of logical analysis.

> A proposition is completely logically analysed if its grammar is made completely clear: no matter what idiom it may be written or expressed in.[25]

But even if Wittgenstein's end is still logical analysis, his means of reaching it have changed.

> I do not now have phenomenological language, or "primary language" as I used to call it, in mind as my goal. I no longer hold it to be necessary [variant: possible]. All that is possible and necessary is to separate what is essential from what is inessential in *our* language.
>
> That is, if we so to speak describe the class of languages which serve their purpose, then in so doing we have shown what is essential to them and given an immediate representation of immediate experience.
>
> Each time I say that instead of such and such a representation you could also use this other one, we take a step towards the goal of grasping the essence of what is represented.
>
> A recognition of what is essential and what inessential in our language if it is to represent, a recognition of which parts of our language are wheels turning idly, does the same job as the construction of phenomenological language.[26]

[24] loc. cit.

[25] *Philosophical Remarks*, p. 51.

[26] *Philosophical Remarks*, p. 51.

The linguistic ontology on which this new methodology rests is so different from Wittgenstein's earlier one (which he shared largely with Russell) and perhaps also from the ones we customarily think in terms of that its specific character is easily overlooked. The Russell who was part of Wittgenstein's background assumed (and argued) that the given, that is, the world of phenomenological objects, can be known, referred to and spoken of unproblematically. It is what is behind the veil of acquaintance, the world of physics, that is not only hard to know but hard even to talk about. Objects of description must be considered as logical constructions out of the given phenomenological entities. Our talk about them is relative to some such logical construction.

In contrast, for Wittgenstein it is the world of everyday physical objects that is semantically primary. It must be dealt with as if it were unproblematic. As we saw, it is for instance only by a dangerous modification of meaning that we can according to Wittgenstein extend many philosophically crucial expressions from their primary naïve physicalistic meaning to the phenomenological realm. As far as language is concerned, it is this world of the immediately given that is hidden behind the curtain of inexpressibility. This semantical hiddenness of the given is one of the features of Wittgenstein's later philosophy that has been for most philosophers too paradoxical to be believed.

One thing that follows from this is that when Wittgenstein criticizes the illusion of a clear logical structure underlying our language, he is not speaking of the world of phenomenological objects and of our actual discourse which is inevitably one of many possible ones. What Wittgenstein says is fully compatible with the idea that "the world we live in" has a clear-cut logical structure, quite as fully as the world of the *Tractatus*. Unfortunately, Wittgenstein no longer believes that that crystalline structure can be mirrored in language. We can speak of it only indirectly, by means of suitable man-made language-games.

In particular, what depends on our language-games and on our "way of life" is the way we speak of the world of physical objects. The phenomenological world is literally given to us. It is independent of us, and therefore the door to the mystical. In the *Tractatus*, it was the existence (the givenness) of the simple phenomenological objects that was the mystical. (See 6.44 and compare it with 5.552.) In this respect, the crash of the philosophy of the *Tractatus* in 1929 does not imply any changes in Wittgenstein's thinking.

Wittgenstein returns to the problems that prompted him to change his mind about phenomenological languages. In *Ludwig Wittgenstein and the Vienna Circle*, p. 97 he is quoted as saying the following:

> How do I verify the proposition "This is yellow"? [...]
>
> An image of "yellow" is not a picture of a yellow that I have seen in the sense in which I carry a picture of my friend, for instance, in my wallet. An image is a picture in an entirely different, formal sense. I may say, "Imagine a certain yellow; now make it become whitish

> until it is completely white, and now make it turn green." By means of this I can guide your images, and they change in the same way as real colour impressions. All operations that correspond to reality I can perform on images. *An image of a colour has the same multiplicity as the colour.* That is what its connection with reality consists in.

Among other things this view presupposes that we can compare the image or picture of what is being searched for not only with the object that the search eventually produces but with any state of affairs, including the current one.

The sentences or pictures that guide one's search must not refer only to the objects and the states of affairs which verify or falsify them. They must be able to guide our search for those objects. They must therefore be capable of being compared with reality at any time.

> It is essential that I must be able to compare an expectation not only with what is considered as the definitive answer (verification or falsification) but also with the present state objects. It is only in this way that an expectation becomes a picture.
> That is, it must have a meaning *now*. (MS 107, p. 284, February 4, 1930.)

This means that the images ("pictures") that we compare with reality must persist in time during the entire search. This provides Wittgenstein with another reason why the objects that we speak of in language cannot be phenomenological, for phenomenological entities do not stay put in the actual physical time which is needed for the search.

But here we can see a difficulty that eventually led Wittgenstein to develop his views in a new direction. The kind of account he is giving still relies on phenomenological entities, viz. images. He can be happy with his new point of view only when he has eliminated such entities from the use of language. We can of course imagine a language community in which actual color samples do the job of images. For instance, Wittgenstein occasionally envisages the possibility that human beings would identify and name colors by means of a color-chart where color samples have signs associated with them. And in PI §53 he says:

> We can also imagine such a table's being a tool in the use of the language. Describing a complex is then done like this: the person who describes the complex has a table with him and looks up each element of the complex in it and passes from this to the sign (and the one who is given the description may also use a table to translate it into a picture of coloured squares). This table might be said to take over here the role of memory and association in other cases. (We do not usually carry out the order "Bring me a red flower" by looking up the colour red in a table of colours and then bringing a flower of the colour that we find in the table; but when it is a question of choosing

> or mixing a particular shade of red, we do sometimes make use of a sample or table.)

But this does not seem to be how our language actually works.

The same problem comes up in connection with rules and rule-following. In fact, another consequence of Wittgenstein's new position, which took him a while to realize fully, is that rules cannot any longer be thought of as phenomenological entities that can be captured by introspection, as he once believed. What remains of a rule is now only its explicit symbolic expression.

> Does following a rule mean following an expressed (uttered) rule? (MS 109, p. 229.)

This leads Wittgenstein to his next major problem. How can a mere symbolic formula or any other "dead" physical object determine and even necessitate my behavior in following the rule it codifies? Even the blueprint of a machine does not necessitate its actual behavior. Thus the rule-following problem is implicit in Wittgenstein's switch of his language paradigm.

It has sometimes been said that the development of Wittgenstein's ideas was gradual, not cataclysmic. I cannot agree. What happened to his views in October 1929 must in the light of what has been found here be considered a major doctrinal and methodological earthquake.

Linguistic Regularity

David Pears

What holds together the things to which a general word applies and distinguishes them from other things? The idea behind Wittgenstein's treatment of linguistic regularity is that the answers given to this question by traditional theories like classical realism and nominalism are empty because there is no independent way of identifying either the universal or the specific similarity invoked. Such theories are failed mimics of science and there is no place for them in philosophy. However, it is not so easy to identify the positive content of his treatment. Why do the pupils who are being taught the meanings of general words by examples of their correct application make such outlandish mistakes?

The answer cannot be that they are probable mistakes or that a teacher would need to guard against them in real life. Wittgenstein's point can only be that they are possible mistakes. They are possible because, if the lesson only proceeds by examples, there will always be many different specifications of the meaning of the word that are satisfied by any finite sequence, and so the pupil can always pick a specification that was not intended by the teacher.[1]

So much is familiar. What is not mentioned so often is that Wittgenstein never suggests that outlandish misunderstandings are probable or that their possibility is at all worrying. If the lesson has been well designed with carefully chosen examples, there will be only one natural way of interpreting them—or perhaps there will be minor variations, to be excluded easily by further examples. The way in which children are taught the sequence of cardinal numbers is similar:

> This hangs together with the question, how to continue the series of cardinal numbers. Is there a criterion for the continuation—for a right

[1] See Saul Kripke, *Wittgenstein on Rules and Private Language: An Elementary Exposition* (Oxford: Blackwell, 1982).

E. De Pellegrin (ed.), *Interactive Wittgenstein*, Synthese Library 349, 171
DOI 10.1007/978-1-4020-9909-0_6, © Springer Science+Business Media B.V. 2011

> and a wrong way—except that we do in fact continue them in that way, apart from a few cranks who can be neglected? [LFM, p. 183][2]

He goes on to explain that the confidence with which we ignore the cranks is not based on our *opinion* that they are wrong, but on a consensus of *action*. This consensus is not something that can be assessed for truth or falsehood. It is, rather, a precondition of any assessment for truth or falsehood.

> This has often been said before. And it has often been put in the form of an assertion that the truths of logic are determined by a consensus of opinions. Is this what I am saying? No. There is no *opinion* at all: it is not a question of opinion. They are determined by a consensus of *action*: a consensus of doing the same thing, reacting in the same way. [LFM, p. 184]

Here he is making the point about the "truths" of logic, but the context shows that he is extending it from its home ground, which is the meanings of individual words.

There are two distinct questions that might be asked at this point. What are the advantages of any consensus in the use of a descriptive vocabulary? And what are the advantages of the consensus that actually underpins our descriptive vocabulary? The first question would introduce the importance of communication with other people and with oneself in the past, while the second question would lead to a very different inquiry—perhaps covering the relative advantages of an everyday vocabulary and a scientific one. The same two questions can be asked about different logics. But I am going to ask something that is prior to all those questions: "At what point on the line between subjectivity and objectivity should Wittgenstein's idea be placed? Is it, perhaps, a neo-Kantian compromise?"

Protagoras said that "man is the measure of all things." That would not be shocking if it meant that we use language in the same way that we use rulers, and that the criteria for the application of words are as objective as the coincidence of the length of an object with the distance between two lines on a ruler. But, of course, it means nothing of the sort; it means that man himself is both measurer and measuring instrument. His reactions have taken over the role of the graduating lines on a ruler and they serve as criteria for the application of his words. That is shocking, or, at least, it has shocked many philosophers.

The idea that man is the measure of all things in this second, more radical sense is the key to understanding Wittgenstein's later treatment of meaning. In the *Tractatus* he had argued that a sentence is a picture:[3]

---

[2] *Wittgenstein's Lectures on the Foundations of Mathematics: Cambridge 1939*, ed. C. Diamond (Ithaca: Cornell University Press, 1976) (LFM).

[3] *Tractatus Logico-Philosophicus*, trans. D. Pears and B. McGuinness (London: Routledge, 1961) (TLP).

2.151 Pictorial form is the possibility that things are related to one another in the same way as the elements of the picture.
2.1511 *That* is how a picture is attached to reality: it reaches right out to it.
2.1512 It is laid against reality like a measure.
2.15121 Only the end-points of the graduating lines actually *touch* the object that is to be measured.

This contact is the relation between name and object, and the leading idea of the picture-theory was that it confers a definite sense on the sentence automatically without any further contribution from the speakers who use it. Wittgenstein's later treatment of linguistic regularity is a rejection of this idea. The sense of a sentence is no longer completely determined by a single application of its words. We also need to know how the speaker will apply each word to other things. Isolated ostensive definitions are never enough to fix their meanings. So we track his use of the word on other occasions, and, if anything is going to be like a ruler, it will be his usage rather than something that he uses—his technique rather than his instrument. He is the measure of all things in the radical sense of Protagoras' dictum.

However, disconcertingly, he is also very unlike a ruler. His criteria are often hard to pin down, and it is sometimes impossible to formulate them in further words. In that case he can only demonstrate his use of a word by actually applying it to things, and it is important that he can often do so without the slightest hesitation. That is what Wittgenstein calls "bedrock": there is no possibility of digging deeper and finding the kind of explanation of his technique that is described in the *Tractatus*. Also, to make matters worse, the meanings of words are apt to change gradually but definitely over fairly long periods of time.

This is the background against which Wittgenstein developed his later account of linguistic regularity. It is guided by two maxims. Never exaggerate the stability of regular patterns and never seek to explain what stability they do have by appealing to theories like classical realism or nominalism.

The difference between his earlier and later treatments of meaning is aptly described by him in a conversation with Waismann in 1931:

> I used to believe, for example, that it is the task of philosophy to discover the elementary propositions. [...] The wrong conception of philosophy which I want to object to in this connection is the following, that we can hit upon something that we today cannot yet see, that we can *discover* something wholly new. That is a mistake. The truth of the matter is that we have already got everything and we have go it actually *present*: we need not wait for anything. We make our moves in the grammar of our ordinary language, and this grammar is already

> there. Thus we have already got everything and need not wait for the future.[4]

But though we now have all the material that we need, it is not at all clear how we should use it to explain linguistic regularity. Conversely, the *Tractatus* had told us exactly how to use material that we did not have.

There are three things that make it difficult to see how to use the available material to explain the phenomenon of linguistic regularity. I will identify them now, and then I will try to explain how Wittgenstein saw, and tried to deal with the three difficulties.

First, there is the obvious fact that dictionaries have to use language to explain the meanings of words. Nobody is worried by the use of language to explain non-linguistic matters, but its use to explain linguistic matters is questionable when what is needed is specifications of meanings.

Second, it is easy to exaggerate the phenomenon of linguistic regularity. The meanings of words develop over time and the changes do not immediately lead to the verdict that there are now two words distinguished only by their meanings.

Third, if we try to demonstrate the meaning of a word by applying it to a series of things, there will always be more than one interpretation of what we have done when we stop at any particular point in the series.[5]

The first problem is an obvious one but its effects are elusive. Evidently, explanations of the meanings of words, like explanations of anything else, will themselves be expressed in words. But why does that present a problem? We do not find fault with a cookery book because it only contains recipes. So why should we complain when a philosophical text contains nothing but verbal analyses of the meanings of words?

We might fail to be reassured by this answer because there is a fundamental difference between the two cases. If we want to know how to follow the recipes in a cookery book, we can ask someone to describe the requisite techniques and we would not complain that all that he gave us was more words. We would be content to hear the techniques described, because we would already know the meanings of the words used in their descriptions. We could, of course, ask for a demonstration, but we would not necessarily feel that the lesson was incomplete without one. But we might well complain if a philosophical text contained nothing but verbal analyses of the meanings of words. We might point out that it omitted an essential part of any theory of meaning: it told us a lot about the connections between words and other words but nothing about the connections between words and the things to which we apply them—nothing about the interface between language and the world.

---

[4] *Ludwig Wittgenstein and the Vienna Circle, Conversations Recorded by Friedrich Waismann*, ed. Brian McGuinness, trans. Joachim Schulte and Brian McGuinness (Oxford: Basil Blackwell, 1979), pp. 182–183.

[5] See Kripke's *Wittgenstein on Rules and Private Language* (op. cit.).

The short answer to this is that, if we want to be told, rather than shown, what happens at that interface, we can only be given more words. But there is a distinction to be drawn at this point: what we are asking for may not be the continuation of the analyses of particular words: it may be a general account of the application of words to things.

This general account is what Wittgenstein offers in his treatment of linguistic regularity in *Philosophical Investigations*. In the much-discussed passage running from §198 to §202 he argues that theories of meaning that merely offer verbal analyses of particular words leave the whole canopy of language "hanging in the air", unattached to anything in our lives. The remedy is to tie it down and that is done in real life by actually applying words to things and in philosophy by giving a general account of the practice of applying words to things.

This part of *Philosophical Investigations* is an implicit criticism of the programme of the Vienna Circle and of the propositions of the *Tractatus* that had inspired it

> 3.26 A name cannot be dissected any further by means of a definition: it is a primitive sign.
>
> 3.261[1] Every sign that has a definition signifies *via* the signs that serve to define it; and the definitions point the way.

The logical atomism of the *Tractatus* had long been abandoned and what he is criticizing in the *Philosophical Investigations*, §§198–202, is preoccupation with definitions and the idea that, when they are no longer available, because he has reached words that are indefinable, all that a philosopher can do is to invoke the doctrine of showing, and say that an elementary proposition shows its sense (TLP 4.22). In fact, that is not its only resource: he can also describe what happens at the interface between language and the world in a way that will make it intelligible. The description can be realistic instead of being driven by the dogmatic fantasies of the *Tractatus.*

The second problem introduced above was that the meanings of words change over time and any philosophical account of linguistic regularity has to allow for that fact. To use Wittgenstein's analogy, the beginning of a series of applications of a word may be like a "visible section of rails" (PI I, §218), but it would be rash to suppose that they can simply be projected unchanged into the infinite future. We have to allow for the vitality and plasticity of language, and the linguistic regularity that we need to understand only extends over limited periods of time. Dictionaries mark the mutations by identifying two words where there had been only one before. During a period of regularity there may be changes, but they must not be so rapid or so extreme that communication between contemporaries breaks down or the recovery of one's own past experiences, preserved in verbal memories, requires scholarship.

The third problem is much more difficult and it is the one that has been most frequently discussed by commentators on Wittgenstein's treatment of linguistic regularity. It arises out of the second problem in the following way: If I have no idea which specification of the series of things to which I have already applied a word ought to be projected into the future, then I do not know what meaning the word will have the next time I use it. I will know the answer to that question only when I have encountered the next thing and decided whether or not the word is applicable to it. So the meaning is dependent on an investigation of my own immediate future: Will I or won't I apply the word to the next problematic thing? But the investigation is unintelligible. For if the word does not take its past meaning with it into the future, the question, whether or not it applies to the next thing, will not have a definite sense. Humpty Dumpty said, "It's my word and I can do what I like with it." But if the paradox of investigation-dependence is validly constructed, how could he even know when he was innovating? The very distinction between continuing a tradition and modifying it seems to have collapsed.

Classical realists will claim that their theory offers the only way of avoiding this collapse. But Wittgenstein rejected their solution to the problem as well as the solution offered by the traditional alternative, nominalism. For neither of these two rival theories gives us anything that is identifiable independently of the phenomena that have to be explained. If, on the other hand, we restrict ourselves to what lies on the surface and is empirically accessible, it just does not seem to be true that "we have already got everything that we need." For we seem to be reduced to using a concept of regularity that runs into the paradox of investigation-dependence. The only answer to the question, "Does the word apply to this thing?", will be, "It applies if I find it applicable", and this account of the relation between word and object seems to eliminate the very ideas of judgement and truth.

The paradox is clearly unacceptable but it may be possible to use the concept of "fit" to dismantle it. When a word is applied to a thing, it looks as if there are only two possibilities. One is that the word has to fit the thing and, if it does fit it, the speaker's claim is true. The other possibility is that the word acquires from the thing the standard of fit that will be imposed on the next thing that it encounters. This is what happens when a thing is used to define the meaning of a word ostensively. Evidently, a single application of a word cannot combine these two functions, because the meaning of a word must be fixed before there is any question of truth. This is the point that is exploited by the paradox of investigation-dependence.

However, the point needs qualification when the word is a general word. For the meaning of a general word will already be anchored, albeit not with complete security, in its previous applications, and so it will not be entirely dependent on its next applications, nor then on the application after that. All its applications will be dependent on the past and may influence its meaning in the future like fashions in clothing or styles of art.

That, at least, is what common-sense prompts us to say. But a defender of the paradox would object that no firm ground can be found in past applications of the word. For the whole sequence of previous applications can always be interpreted in many different ways.

Against this, it is worth pointing out that nobody who is actually applying a general word would feel that he has an absolutely free hand at the moment of application. On the contrary, he would say that the technique that had been developed in his previous applications forced him to apply (or withhold) it on this particular occasion.

There are two points to note here. First, what he does on this occasion will help to constitute his technique only on the next occasion of its deployment. If this were not so, his doing what he does on this occasion would be part of his reason for doing what he does on this occasion—which *would* be absurd. This is, in fact, the absurdity presented by the paradox of investigation-dependence. But it can be avoided if we confine ourselves to using earlier manifestations of his technique when we are specifying it on a given occasion of its deployment. *This* application of the word will help to constitute his technique only on the next occasion. The technique works like a zip-fastener: the closing of each link is based on the closing of the previous links in the series and it then contributes to setting the stage for the closing of the next link. It is this contribution that makes innovation a real possibility in the case of language.

The second point that is needed in order to dismantle the paradox is also concerned with the specification of the speaker's technique. What generates the paradox is the assumption that all specifications of his technique that fit his performance up to the present moment of time have an equal claim to consideration. So in discussions of the paradox the future is the playground of fantasy, and though any future will become the present, there will always be another future in which the sceptic's game can be continued. But this simply ignores the point that Wittgenstein makes in *Lectures on the Foundations of Mathematics*: the set of things picked out by the techniques must be one that we human beings find natural.

It may be objected that human nature is shown in what people do and cannot, therefore, be used to explain what people do. Like the universals of classical realism and the similarities of nominalism, it fails to qualify as an independent identifiable factor. This is a valid objection, but what it shows is not that the general appeal to human nature is misguided, but only that it must be followed by a detailed account of our endowment and needs that will explain why we sort things in the ways that we do. This, of course, raises the question whether there is a clear line of demarcation between this part of philosophy and science.[6]

I will end this paper with a brief review of Wittgenstein's treatment of basic naturalness. It does not include a scrutiny of the borderline between philosophy and science. What he does can be divided into two parts. First, he gives a minimalist account of what it is like to continue the regular application of a word. Then he asks what makes us embellish this account by adding to

[6] Cf. PI II, § xii.

it the picture of an irresistible external force exerted on us by something that is completely independent of our natures.

The minimalist account is given succinctly in PI §219:

> "All the steps are really already taken" means: I no longer have any choice. The rule, once stamped with a particular meaning, traces the lines along which it is to be followed through the whole of space.—But if something of this sort really were the case, how would it help?
>
> No; my description only made sense if it was to be understood symbolically.—I should have said: *This is how it strikes me.*
>
> When I obey a rule, I do not choose.
>
> I obey the rule *blindly.*

Why blindly? Because the constraint comes from within—from our own natures—and not from any external force, and so there is nothing to be seen. The idea, that there is a line to be followed through the whole of logical space, must be understood symbolically rather than literally. It is only a dramatization of the experience of continuing the application of a word, or of continuing the development of a mathematical series.

> "We see a series in just *one* way!"—All right, but what is that way? Clearly, we see it algebraically, and as a segment of an expansion. Or is there more in it than that?—"But the way we see it surely gives us everything!"—But that is not an observation about the segment of the series; or about anything that we notice in it; it gives expression to the fact that we look to the rule for instruction and *do something,* without appealing to anything else for guidance. [PI I, §228]

> I believe that I perceive something drawn very fine in a segment of a series, a characteristic design, which only needs the addition of "and so on" in order to reach to infinity. [PI I, §229]

> "The line intimates to me which way I am to go" is only a paraphrase of: it is my *last* arbiter for the way I am to go. [PI I, §230]

> The rule can only seem to me to produce all its consequences in advance if I draw them *as a matter of course.* As much as it is a matter of course for me to call this colour "blue". (Criteria for the fact that something is a "matter of course" for me.) [PI I, §238]

We externalize our feeling that something is a matter of course and it is easy to see how this might generate either of the traditional theories, classical realism or nominalism.

Wittgenstein also has another point to make about this externalization, a point that is not so easy to understand. He uses an analogy to make the point: the mind of a person who is about to develop a series of regular applications by applying a word to yet another thing is like a machine that is seen as a symbol of its own imminent action. This is not a mechanistic theory of mind but only an analogy used by Wittgenstein to illustrate the origin of a philosophical illusion.

The illusion is a misunderstanding of the experience of "grasping the whole use of a word in a flash." The misunderstanding has two causes. First, it feels to us as if the future applications of a word have already been made and are somehow stored in our minds for future use. Second, we conflate the two kinds of necessity, logical and nomological, and imagine that we can deduce this part of our future with a necessity that combines the strength of logical necessity with the informativeness of nomological necessity. (There is a second, more difficult application of these ideas in PI §§139–140, where Wittgenstein explores the roles played by logical compulsion and psychological compulsion in the interpretation of a picture.)

The first of the two illusions is analysed in PI §§187–188:

> "But I already knew, at the time when I gave the order, that he ought to write 1002 after 1000."—Certainly; and you can also say you *meant* it then; only you should not let yourself be misled by the grammar of the words "know" and "mean". For you don't want to say that you thought of the step from 1000 to 1002 at that time—and even if you did think of this step, still you did not think of other ones. When you said "I already knew at the time ..." that meant something like: "If I had then been asked what number should be written after 1000, I should have replied '1002'." And that I don't doubt. This assumption is rather of the same kind as: "If he had fallen into the water then, I should have jumped in after him."—Now, what was wrong with your idea? [PI I, §187]

It is easy to see how this illusion works: We cannot cross a real bridge until we come to it,[7] but we can always substitute an imaginary bridge in our minds and cross it now. So we forget that the real problem about the development of a technique is the future.

The second illusion is more complex and more difficult to understand. How can a machine be taken as a symbol of its own future action? And how is that way of seeing a machine related to the conflation of the two kinds of necessity? And what has any of this to do with an illusion which generates a theory like classical realism?

> "It is as if we could grasp the whole use of the word in a flash." Like *what* e.g.?—*Can't* the use—in a certain sense—be grasped in a flash? And in *what* sense can it not?—The point is, that it is as if we could 'grasp it in a flash' in yet another and much more direct

[7] See *Wittgenstein's Lectures: Cambridge, 1930–1932. From the Notes of John King and Desmond Lee*, ed. Desmond Lee (Oxford: Basil Blackwell), p. 67.

> sense than that.—But have you a model for this? No. It is just that this expression suggests itself to us. As the result of the crossing of different pictures. [PI, §191]

> You have no model of this superlative fact, but you are seduced into using a super-expression. (It might be called a philosophical superlative.) [PI, §192]

> The machine as symbolizing its action: the action of a machine—I might say at first—seems to be there in it from the start. What does that mean?—If we know the machine, everything else, that is its movement, seems to be already completely determined.
> [...] [PI I, §193]

We need an example of a machine and the ways in which we see it. So picture a clock with its mechanism encased in glass, so that the arrangement of its working parts is clearly visible. Evidently, there will be two ways of looking at this clock. On the one hand, someone with an elementary knowledge of mechanics but no knowledge of clocks might see how it would work if it were wound up. On the other hand, another observer, with no knowledge of mechanics, might see how it would work because he is familiar with the function of clocks. The first observer would be relying on nomological necessity, while the second observer would be relying on logical necessity—if this is a clock, it will function as its maker intended it to function.

If the clock malfunctions, both these observers will be criticized for drawing an exaggerated conclusion from a necessity. The first one forgot that his prediction was not nomologically necessary because he ought to have inserted conditions, such as, "if the spring does not break" or "if the balance-wheel does not jam." The second observer forgot that his prediction was not logically necessary, because he ought to have inserted the condition, "if all goes according to the maker's plan" (which, of course, summarizes the conditions omitted by the first observer).

It is an important feature of both these observers' mistakes that they are caused by a failure to allow for mechanical breakdown, which is, of course, the analogue of applying a word in a way that is not the right projection of its past applications. But the analogy is imperfect. For, as I mentioned earlier, innovative applications of a word do occur and, if they catch on, they do not immediately make us distinguish two words where there had been only one before. But we do not look at clocks in this way. If I returned a faulty clock to the shop where I had bought it, I would not expect to be told that a clock is a clock, and if this one will not tell the time correctly, perhaps I could find some other domestic use for it.

It is the second observer who sees the machine as a symbol of its own future performance. Like the first observer, he forgets the things that may go wrong and reads its future performance in the present state. But he also converts the first observer's nomological necessity into a logical necessity and so achieves an illusory hybridization.

On a Remark by Jukundus

Joachim Schulte

1. In 1946, in a note written down in the context of his work on the philosophy of psychology, Wittgenstein said:

> The remark by Jukundus in 'The Lost Laugh', that his religion consisted in: his knowing, if things are going well for him now, that his fate could take a turn for the worse—this actually is an expression of the same religion as the saying "The Lord hath given, the Lord hath taken away."[1]

Jukundus is the protagonist of a story by Gottfried Keller, who was one of Wittgenstein's favourite writers. The title of the story is "*Das verlorene Lachen*" ("The Lost Laugh"), and in the passage Wittgenstein presumably had in mind when jotting down the above-quoted remark Jukundus speaks the following words:

> I do believe that in point of fact I know what it means to fear God, in that as regards fate and life I am quite incapable of saying anything irreverent. I do not think I have a right to ask that everywhere and as a matter of course things should go well; rather, I am afraid that here and there things may take a turn for the worse [...].[2]

---

[1] *Culture and Value* (C&V), ed. Georg Henrik von Wright in collaboration with Heikki Nyman, revised edition of the text by Alois Pichler, trans. Peter Winch (Oxford: Blackwell, 1998) p. 54. References are to the relevant pages of this edition. ["Die Bemerkung des Jukundus im 'Verlorenen Lachen', seine Religion bestünde darin: er wisse, wenn es ihm jetzt gut geht, sein Schicksal könne sich zum Schlechten wenden—dies drückt eigentlich die gleiche Religion aus wie das Wort "Der Herr hat's gegeben, der Herr hat's genommen."]

[2] Gottfried Keller, "Das verlorene Lachen", in *Sämtliche Werke*, ed. Jonas Fränkel (Erlenbach-Zürich and München: Eugen Rentsch, 1927) p. 345. ["Ich glaube, der Sache nach habe ich wohl etwas wie Gottesfurcht, indem ich Schicksal und Leben gegenüber keine Frechheit zu äussern fähig bin. Ich glaube nicht verlangen zu können, dass es überall und selbstverständlich gut gehe, sondern fürchte, dass

E. De Pellegrin (ed.), *Interactive Wittgenstein*, Synthese Library 349,
DOI 10.1007/978-1-4020-9909-0_7,

Many readers of the collection *Culture and Value* must have wondered about Wittgenstein's remark and the meaning of the purported equation between the fear of God expressed by Jukundus and the religious attitude of Job. Does the quoted note have any kind of deeper significance at all?

In this paper I wish to explain and defend the thesis that, if the above-quoted note is read the right way and seen in connection with certain other considerations of Wittgenstein's, it really does serve to illuminate and summarize an important strand of his thinking. To make this thesis clearer and more persuasive, however, I shall have to supply some background.

2. There are, as a matter of fact, a fair number of books and articles on Wittgenstein's real or presumed contributions to philosophy of religion, but from the very beginning it should be unambiguously clear that he never saw himself as the author of writings belonging to the philosophy of religion. To be sure, there are students' notes of three "Lectures on Religious Belief", but notes of this kind can hardly ever be particularly reliable sources. And in addition, the third of these lectures is chiefly devoted to spiritualism and related questions. The most important documents are Wittgenstein's own notes, as far as they are extant among his *Nachlass* writings. Besides these, there are instructive reports of discussions with friends or pupils. A fair number of pertinent remarks is contained in the collection *Culture and Value* and in the edited diary *Denkbewegungen.*[3]

The relevant remarks extant among Wittgenstein's *Nachlass* writings are scattered and should not be regarded as the makings of a "philosophical theory of religion." Further interesting material is contained in "Remarks on Frazer's *Golden Bough*", the first part of which was written in 1931 and partly re-ordered by Wittgenstein himself. (The equally interesting second part of these "Remarks" was written very much later on loose sheets of paper. After Wittgenstein's death these sheets were found in his one-volume copy of the *Golden Bough.*)[4]

---

es hie und da schlimm ablaufen könne."] As far as Wittgenstein is concerned, only the above-quoted words are to the point. As regards Keller (and Jukundus), however, the sequel too is significant: "[...] and I hope that things will none the less take a turn for the better." ["und hoffe, dass es sich dann doch zum Bessern wenden werde"]

[3] Wittgenstein, *Lectures and Conversations on Aesthetics, Psychology and Religious Belief*, ed. Cyril Barrett (Oxford: Blackwell, 1966); *Denkbewegungen: Tagebücher 1930–1932, 1936–1937* (DB), ed. Ilse Somavilla (Innsbruck: Haymon, 1997). (An English translation of DB by Alfred Nordmann is in preparation; Nordmann has kindly given me permission to quote from his translation.) In the case of DB references are to the pages of the original manuscript (these pages are given in the printed edition of DB). Occasionally I depart from published or unpublished translations without alerting the reader to the fact that I am doing so.

[4] Wittgenstein, "Bemerkungen über Frazers *The Golden Bough*" (Fr), ed. Rush Rhees, in *Synthese* (1967), pp. 233–253. Revised reprint (with an English transla-

On various occasions Wittgenstein displayed an interest in the phenomena of religious life. It is well-known that as a young man he read and expressed his appreciation of William James's *Varieties of Religious Experience.*[5] Later he got interested in Frazer's *Golden Bough* and concerned himself with works by Ernest Renan, Karl Barth and no doubt other authors belonging to this field. Wittgenstein loved certain decidedly Christian authors such as (of course!) Augustine, Angelus Silesius, Matthias Claudius as well as writers in whose works religious themes play a great role (e.g. Tolstoy and Dostoevsky). He read and admired some of Lessing's writings on theological questions; and at least for a certain time Kierkegaard played an important part in his thinking.

Remarks on what might be regarded as religious questions in a wide sense can be found in Wittgenstein's writings from various periods of his life.[6] There is a striking concentration of such remarks in the years 1936 and 1937. That was a particularly critical time for Wittgenstein. He had left Cambridge and moved to Norway to get something written. He discarded his first effort (an attempt at a revised German version of the *Brown Book*) and started something new (something which much later became the first third of *Philosophical Investigations*). He had not yet managed to settle accounts with himself and now forced himself to confess his "sins" to friends and relatives.[7] Sometimes the loneliness of his Norwegian life was hard to bear. No wonder that his notebooks and diaries from this time contain not only general remarks on religious topics but also notes that one might take as expressions of a "struggle for consolation" or a "struggle with God."

---

tion by John Beversluis) in *Philosophical Occasions: 1912–1951*, ed. James Klagge and Alfred Nordmann (Indianapolis: Hackett, 1993), pp. 115–155. References are to pages of the first edition, which are reproduced in the later edition. Also relevant are certain passages in Friedrich Waismann's notes of conversations with Wittgenstein (*Ludwig Wittgenstein and the Vienna Circle* (WVC), ed. Brian McGuinness, trans. Joachim Schulte and Brian McGuinness (Oxford: Blackwell, 1979)) as well as Wittgenstein's "Lecture on Ethics" (LoE), reprinted in *Philosophical Occasions*, pp. 37–44.

5 See letter to Russell, 22 June, 1912, in *Cambridge Letters*, ed. Brian McGuinness and Georg Henrik von Wright (Oxford: Blackwell, 1995), p. 14.

6 For the early Wittgenstein, see Brian McGuinness, "The Mysticism of the Tractatus", in *Philosophical Review* (1966), pp. 305–328; reprinted (under the title "Mysticism") in McGuinness, *Approaches to Wittgenstein: Collected Papers* (London and New York: Routledge, 2002), pp. 140–159.

7 See Fania Pascal, "Wittgenstein: A Personal Memoir", in *Recollections of Wittgenstein*, ed. Rush Rhees (Oxford: OUP, 2nd edition 1984), pp. 12–49, especially 34–39 ("The Confession"); Ilse Somavilla, Anton Unterkircher, Christian Paul Berger (eds.), *Ludwig Hänsel – Ludwig Wittgenstein: Eine Freundschaft* (Innsbruck: Haymon, 1994), letters 225–229; Brian McGuinness, Maria Concetta Ascher, Otto Pfersmann (eds.), *Familienbriefe* (Wien: Hölder-Pichler-Tempsky, 1996), letter 123; *Cambridge Letters*, letter 169; Paul Engelmann, *Letters from Ludwig Wittgenstein with a Memoir*, ed. Brian McGuinness, trans. L. Furtmüller (Oxford: Blackwell, 1967), letter 54; DB, 124, 146.

But never was Wittgenstein a religious or devout man in any ordinary sense of these words. He himself might have said that he was incapable of taking the crucial step in the direction of religiousness, for as he saw it he could not kneel down to pray.[8] Certain aspects of religious life were completely alien to him. When a former pupil of his converted to Catholicism Wittgenstein wrote to him and said: "If someone tells me he has bought the outfit of a tightrope-walker[9] I am not impressed until I see what is done with it."[10]

If one disregards Wittgenstein's references to the religious practices described in Frazer's book, his remarks on religious questions or phenomena generally refer to Christianity.[11] Here two points are striking: First, for Wittgenstein genuine religiousness is always connected with decisions on how to lead a decent life. One might say that his view of religion was a profoundly ethical one (which at the same time throws a particular light on his idea of "ethics").[12]

---

[8] C&V, 63 (cf. DB, 183, 210). This remark, however, must not be taken literally. It may well be that occasionally Wittgenstein knelt down to pray (cf. DB, 184), but what was not even remotely possible for him was due obedience to God. Anticipating what I shall say in §10 below, one might claim that while Wittgenstein was quite capable of kneeling down to pray in order to console or to humiliate himself, etc., he was not able to do so for the purpose of making a *gesture of deference, submission, or subordination.*

[9] Evidently the comparison between a religious person and a tightrope-walker came naturally to Wittgenstein. Cf. C&V, 84: "The honest religious thinker is like a tightrope walker. It almost looks as though he were walking on nothing but air. His support is the slenderest imaginable. And yet it really is possible to walk on it." Cf. also Ulrich Arnswald and Anja Weiberg (eds.), *Der Denker als Seiltänzer: Ludwig Wittgenstein über Religion, Mystik und Ethik* (Düsseldorf: Parerga, 2001).

[10] Drury, "Some Notes on Conversations with Wittgenstein", in Rhees (ed.), *Recollections of Wittgenstein*, 88. An even clearer expression of Wittgenstein's attitude is the following passage (DB, 69 f.): "That nowadays someone would convert from Catholicism to Protestantism or from Protestantism to Catholicism is embarrassing to me [...]. Something that can (now) make sense only as a tradition is changed like a conviction. It is as if someone wanted to exchange the burial rites of our country for those of another.—Anyone converting from Protestantism to Catholicism appears like a mental monstrosity. No good Catholic priest would have done that, had he been born a non-Catholic. And the reverse conversion reveals abysmal stupidity. | Perhaps the former proves a deeper, the latter a more shallow stupidity."

[11] Wittgenstein's notorious remarks on Jews have no connection with religious aspects of Jewishness. For this topic, see McGuinness, "Wittgenstein and the Idea of Jewishness", in James C. Klagge (ed.), *Wittgenstein: Biography and Philosophy* (Cambridge: Cambridge University Press, 2001), pp. 221–236 (under the title "The Idea of Jewishness" in *Approaches to Wittgenstein*, 27–42); David Stern, "Was Wittgenstein a Jew?", in Klagge (ed.), op. cit., pp. 237–272.

[12] Wittgenstein abhorred ethical "theories." If something is to deserve the title "ethics" it must essentially be pronounced in the first person—if there is anything to be pronounced at all. See the following well-known passage from WVC, 117: "If I could explain the essence of the ethical only by means of a theory, then

Second, the number of religious doctrines and practices mentioned again and again is extremely small. What he discusses are the notion of religious belief, the question of election by grace (predestination), the idea of God as judge, confession, and prayer. Even if it appears that he himself is strongly affected by a given question he seems quite incapable of shedding his "ethnological" way of looking at things. He remains true to his maxim expressed in the following passage:

> If we use the ethnological approach does that mean we are saying philosophy is ethnology? No, it only means we are taking up our position far outside, in order to see things more *objectively.* (C&V, 45)

3. According to Wittgenstein, questions of religion and questions of life (practice) cannot really be separated.[13] Both religion and life stand opposite the province of theory. Theology, or certain forms of theology, do not lead to clarity. That is not because the authors of theological writings express themselves obscurely; rather, the reason is that in religion words do not matter—at any rate they do not matter in the sense generally relevant to theology.[14] "Theology that insists on *certain* words and phrases and prohibits others makes nothing clearer. (Karl Barth) | It brandishes [*fuchtelt*] words, as it were, because it wants to say something and does not know how to express it. *Practice* gives the words their sense." (C&V, 97) This remark is to be taken quite literally: while there are many linguistic expressions that may obtain their senses by means of theoretical instruction, in religion one finds, according to Wittgenstein, words (or certain uses of words) that outside religious contexts have neither meaning nor an intelligible explanation.

At no point does Wittgenstein speak of religion in terms of a revealed doctrine or an insight into transcendent matters.[15] He quotes a passage from Lessing (C&V, 11), where this author describes the style of the Bible and

---

what is ethical would be of no value whatsoever. | At the end of my lecture on ethics I spoke in the first person: I think that this is something very essential. Here there is nothing to be stated any more; all I can do is to step forth as an individual and speak in the first person." Cf. LoE, 44, and DB, 76: "But an ethical proposition is a personal act."

[13] "Amongst other things Christianity says, I believe, that sound doctrines are all useless. That you have to change your life. (Or the direction of your life.)" (C&V, 61)

[14] Of course, there is another sense in which words matter a great deal in religion. That is for instance the case in situations where the articulation of powerful images is important.

[15] This fact annoys readers like John W. Cook: "[W]e can look at the creeds orthodox Christians recited each Sunday. When people recite such creeds they are not philosophising, they are reciting what they believe. Only an a priori theory like Wittgenstein's would lead one to think otherwise. [...] Wittgenstein would have a problem with the idea of divine agency just because the cause is transcendent, not given in experience. [...] Had he [Wittgenstein] remained true to his claim to be replacing wild conjectures with quiet weighing of linguistic facts, he would have

mentions the tautological and deceptive aspects of the book. Again in a context of discussing Lessing, Wittgenstein wonders what it means to claim that when confronted with the Bible a man has "nothing but a book" before him (DB, 148f.). As such it really does not have a different status from other documents. It will become binding only if it proves evident in a particular way; and it is not only as an ethical work that it will have to achieve this degree of evidence but also as an historical account. This is obviously a point where faith will have to play a decisive role; you will have to believe things (in a special use of the word "believe").

Wittgenstein writes that Christianity is "not a doctrine." (C&V, 32) It is not a theory about this or that. In a way it even stands in opposition to intelligence [*Klugheit*] and wisdom: "Christianity is really saying: let go of all intelligence." (DB, 130) "'Wisdom is grey.' Life on the other hand and religion are full of colour." (C&V, 71) It would be a misreading of these passages if one wanted to explain them by suggesting that they had to be understood in the sense of "pedantic" or "hair-splitting intelligence", in the sense of "pompous" or "schoolmasterly wisdom." No, as Wittgenstein sees it, it is quite possible to read "intelligence" and "wisdom" in a positive sense while seeing them as standing in fundamental opposition to religion. Why does he think so? A first clue may be derived from the following remark:

> Wisdom is something cold, and to that extent foolish. (Faith, on the other hand, a passion.) We might also say: wisdom merely conceals life from you. (Wisdom is like cold, grey ash covering the glowing embers.) (C&V, 64)

In this passage Wittgenstein alludes to an idea of Kierkegaard's which he discusses in another manuscript (C&V, 61). Kierkegaard's idea is that faith is a passion.[16] That is the reason why one can be "seized" and "turned around" by faith. Wisdom, on the other hand, is something one can follow in a cool and collected way "like a doctor's prescription." In Wittgenstein this idea is connected with the contrast "cold/hot." Wisdom is cold. Its council may be helpful and thus turn out to provide the best guidance in most situations of life. But it is no use when it is a question of "setting your life to rights, [no more] than you can forge iron when it is *cold*." A man who wishes to change his life or cannot help attempting to do so needs religion or something of the kind of religion—at any rate, he will need a hot, a "passionate" element to re-forge his life.

opted [...] for the [...] alternative, that religious belief is nonsense" ("Religious Belief", in *Wittgenstein's Intentions*, ed. John V. Canfield, Stuart G. Shanker (New York and London: Garland, 1993), pp. 147–161, quotations pp. 156–158).

16 In the context of religious belief it seems quite natural to talk of passion: one can adduce examples to show what is meant by speaking this way. On the other hand, if someone *passionately* utters the words "I know that this is a foot" his passion should rouse our suspicion (cf. Wittgenstein, *On Certainty*, §§376–380).

In his remarks on Frazer Wittgenstein discusses another thought that is connected with this idea of passion, viz. the notion of ceremony or ceremoniousness. This is an aspect of religion in general. Ceremony is not something normal or casual. It is exaggerated, excessive, extreme, like the expression of a passion. The example Wittgenstein mentions is characteristic of his way of thinking: After Schubert's death his brother cut small fragments from some of his scores and gave them as tokens to friends of the composer. That was not a casual act; it was a reverent sort of behaviour, a kind of ceremony. As an expression of awe or reverence (*Pietät*) it is just as intelligible to us as burning or locking up the scores would have been. In any event, ceremony is not without passion, it is not lukewarm but either hot or cold: it is something out of the ordinary.[17]

4. Certain themes known from Wittgenstein's philosophical writings can be seen to play a role in his religious reflections too. Thus in the *Tractatus* Wittgenstein writes: "The sense of the world must lie outside the world. In the world everything is as it is, and everything happens as it does happen: *in* it no value exists.—And if it did exist, it would have no value." (6.41) Of this one may feel reminded when reading a remark from 1929 where Wittgenstein says that what is good is divine too. And that idea, he observes, is an epitome of his ethics. Only what is supernatural—and hence no statement about the facts of this world—can express the supernatural (C&V, 5).[18]

Another remark Wittgenstein makes in the *Tractatus* is that the solution of all possible scientific questions leaves our problems of life completely untouched (6.52). Much later he notes that the problems of life cannot be solved in surface dimensions but, if at all, by moving into certain depths (1948, C&V, 84). Presumably here, too, what is meant by "surface" is the domain of scientifically or empirically discoverable fact. In Wittgenstein's writings the notion of depth, on the other hand, often relates to something instinctive and in this sense to something "primitive" (cf. especially Section II of the remarks on Frazer, and see below, note 1 of the Appendix). By no means does he want to suggest that the solution of the problems of life is to be looked for in a transcendent sphere. While Wittgenstein recognizes that there are religious, e.g. Christian, solutions of such problems, even these solutions are never connected with something "transcendent" but with certain images that come to the fore in specific ways of life. Through an encounter with such images one may feel forced to adopt a certain attitude towards these images,[19] and in this way it may come about that the world appears in a new light; but that

[17] "The ceremonial (hot or cold) as opposed to the haphazard (lukewarm) characterizes reverence." (Fr, 238)

[18] Cf. DB, 25f.: "[...] genuine nimbus does not attach to external fact, that is, not to facts."

[19] "Believing in an apostle means to relate toward him in such and such a way—relate actively" (DB, 74). "On Kierkegaard: I represent a life for you and now see how you relate to it, whether it tempts (urges) you to live like that as well, or

does not necessarily involve changing one's *opinions*.[20] That, however, does not mean that nothing changes; after all, "if one lives differently, one speaks differently. With a new life one learns new language games." (DB, 161) And the fact that one is dealing with problems or questions of life is a crucial mark of the religious domain:

> A *religious question* is either a question of life or it is (empty) chatter. This language game—one could say—gets played only with questions of life. Much like the word "ouch" does not have any meaning—except as a scream of pain. — I want to say: If eternal salvation means nothing for my life, *my way of life*, then I ought not to rack my brain about it; if I have the right to worry about it, then what I think must stand in a precise relation to my life, otherwise what I think is rubbish or my life is in danger. (DB, 203f.)

Here as elsewhere in Wittgenstein, the reader should be careful when dealing with the concept "language game." To be sure, there are boundaries between different language games, but it is not always possible to draw these boundaries in a sharp way, nor need the boundaries stay the same if you change your point of view. In writing that a "religious language game" should, roughly speaking, concern nothing but questions of life Wittgenstein does not make a neutral and in *this* sense "descriptive" statement according to which there exists, besides various other kinds of language games, a clearly delimited language game of religion. Obviously, what he says is an expression of his personal appraisal of the matter; and it is *his* view that only someone who takes religion so seriously is to count as a really religious person.

5. Modifying a quotation from Schopenhauer, Wittgenstein says in a conversation of 1930: "To moralize is difficult, to establish morality impossible." (WVC, 118)[21] Similar remarks can be found among Wittgenstein's notes on religious precepts or maxims. What is particularly instructive are the following ideas which he develops in the context of his reflections on religious similes: In *The Pilgrim's Progress*, Wittgenstein says, Bunyan's similes remain unconvincing because they suggest a continuation that would not really fit the

what other relation to it you attain. Through this representation I should, as it were, like to loosen up your life." (DB, 75)

[20] This is based on a remark of Wittgenstein's written in 1937 (DB, 161). It is likely that "opinions" is to be understood in an analogous way (but, of course, only in an *analogous* way) to "propositions with a sense" in the *Tractatus*. Just as the good or bad exercise of the will, happiness or unhappiness cannot change the facts—i.e. that which "can be expressed by language" (6.43)—so in this later remark too "opinions" remain independent of the "images" of a "new life."

[21] Cf. DB, 76: "Just consider that the justification of an 'ethical proposition' merely attempts to refer the proposition back to others that make an impression on you. If in the end you don't have disgust for this and admiration for that, then there is no justification worthy of that name."

intended cases. Too easily one might conclude that all the traps into which a man in quest of religion may fall have been laid by God himself. And as soon as this suspicion comes into view those similes lose whatever persuasiveness they may have had until then. This weakness affects them because they, *qua* similes, are here used to justify certain claims. It would be different if they were simply employed as pictures to characterize prescribed or forbidden forms of conduct, that is, if they were used without any pretensions to justification:

> Rules of life are dressed up in pictures. And these pictures can only serve to describe what we are supposed to do, but not to justify it. Because to be a justification they would have to hold good in other respects too. I can say: "Thank these bees for their honey as though they were good people who have prepared it for you"; that is intelligible and describes how I wish you to behave. But not: "Thank them, for look how good they are!"—since the next moment they may sting you. (C&V, 34)

Similarly, Wittgenstein thinks, religion can only convince us if it uses pictures that simply serve to describe something. These images may be strange, even repugnant, but they must not suggest the notion that here an attempt is made to justify a claim.[22] For in that way one would enter the level of reasoning and argument, and at that level one might rightly feel challenged to respond and to contradict the claim made or implied. But that would of course be an entirely inappropriate response to what is alleged to be a commandment originating in the will of God.

Thus images or pictures in the sense relevant here are something fundamentally different from similes. Pictures, in contrast with similes, can without further ado force themselves upon you. As was pointed out above, a new life "shifts completely different images into the foreground, *necessitates* completely different images. Just like trouble teaches prayer." (DB, 161) What "can be said by means of a simile, that can also be said without a simile." (DB, 173) In other words, similes can generally be translated; their content can be put differently without losing anything of crucial importance. Pictures or images, on the other hand, can fulfil their functions—that is, they can *be* pictures—only in certain spheres of life. "These images and expressions have a life only in a *high* sphere of life, they can be rightfully used only in this sphere. All I could really do is make a gesture which means something similar to 'unsayable', and say nothing." (ibid.) The fact that Wittgenstein calls the

[22] Cf. C&V, 97: "A proof of God ought really to be something by means of which you can convince yourself of God's existence. But I think that *believers* who offered such proofs wanted to analyse and make a case for their 'belief' with their intellect, although they themselves would never have arrived at belief by way of such proofs. 'Convincing someone of God's existence' is something you might do by means of a certain upbringing, shaping his life in such and such a way."

sphere of life going with those pictures a "high" one is just as characteristic of him as is the underlying observation that *he himself* has no place within this sphere and that the only thing he can do is keep silent and make a gesture expressive of his incapacity to say anything.[23]

The unmistakable *Tractatus* echo is not the only noteworthy feature of this remark, which was written in 1937. What is also significant is that Wittgenstein sees himself as capable of that gesture. After all, it seems that one has the *right* to manifest this specific kind of speechlessness vis-à-vis certain expressions or images only if one commands a minimum degree of understanding of those expressions or images. Is this impression correct?

But the difficulty one may feel one is facing here appears to be purely imaginary. Take the example of Wittgenstein's claim that he for his part cannot assert that God will sit in judgement over him. That need not mean that for him the idea of God as judge is unintelligible nonsense. Of course, what suggests that here we might have a case of nonsense is the (presumed) fact that he would be equally incapable of saying that God will *not* sit in judgement over him. (If it is impossible to assert either p or $\sim p$, one tends to suspect that p is nonsense.) Another possible parallel would be that with "performative" infelicities or suchlike. Thus a 12-year old boy "can" neither ask an adored woman to marry him nor "can" he refuse to marry her—he simply does not belong to the group of admissible suitors or jilters. In a similar way, one might want to suggest, he who does not conceive of God as sitting in judgement over him "cannot say" that God will be his judge.[24]

---

[23] As regards the idea of belonging to higher or lower spheres of life, see the following diary entry of 1931 (DB, 86f.): "Most of the time my justness, when I am just, stems from cowardice. | By the way I don't condemn *that* justness in me which plays itself out on, say, a religious plane onto which I escape from the filthy baseness of my lust and listlessness. This escape is right when it happens out of disgust with that filth. | That is, I am doing right when I proceed to a more spiritual plane on which I can be a human being—while others can be human also on a less spiritual one. | I just don't have the right to live on that floor as they do and on their plane feel my inferiority rightfully. | I must live in a more rarefied atmosphere and belong there; and should resist the temptation of wanting to live in the thicker layer of air with the others, who are allowed to do so."

[24] It may happen that the apparent nonsensicalness of a way of acting serves to draw our attention in a direction which proves helpful. In such a case the apparent nonsensicalness may be due to the (seeming) impossibility to apply the law of non-contradiction or to "performative" aspects of the situation. As Wittgenstein indicates (Fr, 249f.), it may however also be due to an apparently pointless means-end relation: "But what prevents us from assuming that the Beltane Festival has always been celebrated in its present (or very recent) form [i.e. without human sacrifice]? One would like to say: it's too foolish [*sinnlos*] for it to have been invented in this form. Isn't that like my seeing a ruin and saying: that must have been a house at one time, for nobody would have put up such a heap of hewn and irregular stones? And if I were asked: How do you know that? I could only say:

6. The questions posed by these considerations require further exploration, but in order to continue this enquiry it will be helpful to discuss some additional thoughts of Wittgenstein's on religion. An especially important idea is connected with the recurrent metaphor of a ladder and its rungs—another image that is reminiscent of the *Tractatus*:

> In religion it would have to be the case that corresponding to every level of religiosity there is a form of expression that has no sense at a lower level. For those still at the lower level this doctrine, which means something at the higher level, is null and void; it *can* only be understood *wrongly*, and so these words are *not* valid for such a person. (C&V, 37)

Thus what has sense and makes sense at one level of religiosity need not be properly intelligible at another level. The importance of this idea is due to the fact that religion has a lot to do with believing, and to our tendency to assume that only what is (supposedly) intelligible can be believed. If you do not know what it means to call a colour warm or cold you can neither believe nor doubt a sentence like "That's a warm shade of beige." On the other hand, it is of course possible to instruct someone who is not familiar with this way of employing the words "warm" and "cold" in this use by means of adducing examples, so that eventually he will be able to apply these words without outside help.

In the case of religion, matters are not as simple as all that. Mere instruction will not be enough to raise someone who is willing to learn about it up to an higher level of religiosity. But why should he not be able to *understand* what a person standing at another level believes? After all, there are for example lots of books on the religions of our world. Does Wittgenstein want to claim that we are simply incapable of grasping part of what is written there? If we wish to answer this question, we shall need to differentiate more finely and have a closer look at Wittgenstein's own examples.

First it needs to be taken into account that religious writings of the type of the Bible are addressed to everyone. Such writings are meant to be understood by all human beings.[25] This kind of understanding would be impossible if it depended on one's having risen to a particular level of religiosity. But quite apart from the fact that there are, generally speaking, religious traditions of completely different types—e.g. more or less pious legends, consolatory poems,

---

from my experience with people. Indeed, even in places where people actually build ruins, they take the form of collapsed houses."

[25] The variety of possible interpretations of biblical texts contributes to their general intelligibility. That is a point to which Wittgenstein draws attention in the following remark: "The parables of the New Testament leave room for as much depth of interpretation as you like. They are bottomless. | *They* have less style than the first speech of a child. Even a work of supreme art has something that can be called 'style', yes even something that can be called 'mannerism'." (C&V, 43)

hermetic doctrines crying out for exegesis etc.—even the Bible itself is not a unified text. In this sense Wittgenstein writes:

> The old Testament seen as the body without its head; the New T[estament]: the head; the Epistles of the Apostles: the crown on the head. | If I think of the Jewish Bible, the Old Testament on its own, I should like to say: the head is (still) missing from this body. The solution to these problems is missing. The fulfilment of these hopes is missing. But I do not necessarily think of a head as having a *crown*. (C&V, 40)

Even as regards the book which consists of the Old *and* the New Testament and thus forms *one* unitary body Wittgenstein would hold that there are many passages that can speak to a person only if he has reached a certain level of religiosity. The images which are expressed or alluded to are simply set out before us. There are those who remain indifferent to them, and there are others who feel addressed. As *images* or *pictures* they will have a function only for the latter type of person. Yet one will not necessarily want to claim that those who feel indifferent or even repelled cannot make head or tail of them. It does not mean that for such people these images have no sense.

The stories recounting the life of Jesus can become enigmatic only if one tries to interpret them. Only someone who wishes to spell out what the whole narrative means will wonder how it might be possible to respond to contradictions and improbabilities by *believing* them. At this level, the right reply might be that a less questionable story would not have the right effect. Exactly because it is unobjectionable and all too straightforward it might distract from what is decisive, viz. the spirit of the story. Attempting to render intelligible a passage from Kierkegaard, Wittgenstein writes:

> God has *four* people recount the life of the incarnate God, each one differently, and contradicting each other—but can't we say: It is important that this narrative should have more than quite middling historical plausibility, *just so* that this should not be taken as the essential, decisive thing. So that the *letter* should not be believed more strongly than is proper and the *spirit* receive its due. I.e.: What you are supposed to see cannot be communicated even by the best, most accurate, historian; *therefore* a mediocre account suffices, is even to be preferred. For that too can tell you what you are supposed to be told. (Roughly in the way a mediocre stage set can be better than a sophisticated one, painted trees better than real ones,—which distract attention from what matters.) (C&V, 36f.)

7. The Gospel stories are unsophisticated. Understanding these reports does not require much learning or great acumen. In spite of this they may lead to considerable difficulties of understanding—maybe helpful difficulties of understanding. One of these difficulties resides in the historical character of this

tradition. To the extent we are here given an historical report it, on the one hand, is subject to the same requirements of consistency, probability and plausibility as other historical testimonies. On the other hand, we are supposed to encounter these reports with an attitude of imperturbable belief. That seems paradoxical, but according to Wittgenstein it is no paradox. One needs to grasp, however, that Christianity does not *rest* on historical truth. It does contain a message which one cannot really help calling an historical one and it also requires us to receive this message with unshakeable trust, but the kind of belief we are asked to muster is at the same time different from the kind of belief expressed by sentences like "I believe (am convinced) that the Boston Tea Party took place in 1773."

Many would at this point want to object that this is a mere verbal trick: the word "believe" is simply used in two different senses—one time in the sense of holding true a normal, generally falsifiable and perhaps verifiable historical report, another time in the sense of unconditionally holding on—"through thick and thin" (C&V, 37)—to something that is a merely *quasi* historical testimony. Basically, it would be more appropriate to use two different words for these different attitudes.

As a matter of fact, Wittgenstein nearly goes so far as to say exactly that—but only *nearly* so. While he does say that in reality the one kind of believing is a sort of "lovingly seizing on" a certain message (C&V, 38),[26] he does not want to change our use of language, and the reason for that surely is not a general reluctance to indulge in linguistic legislation. Probably he wants to suggest that the interaction between different senses of the word "believe" is something we *need* if we are to be able to say certain things. The claim that the "historical proof-game" (ibid.) cannot begin to play a role when we are dealing with questions of religious belief is an intelligible statement only if there is an echo of the usual meaning of our word "believe." On the other hand, many of our *prima facie* 'normal' uses of the word "believe" are similar to religious uses of that word;[27] and that particularity too becomes clear only

[26] See the parallel formulation in *On Certainty* (§167): "He [Lavoisier] seizes on [grasps—*ergreift*] a given world-picture—not of course one that he invented: he learned it as a child. I say world-picture and not hypothesis, because it is the matter-of-course foundation for his research and as such also goes unmentioned." Cf. C&V, 73: "It appears to me as though a religious belief could only be (something like) passionately committing oneself to a system of coordinates. Hence although it's belief, it is really a way of living, or a way of judging life. Passionately taking up *this* interpretation. And so instructing in a religious belief would have to be portraying, describing that system of reference and at the same time appealing to the conscience. And these together would have to result finally in the one under instruction himself, of his own accord, passionately taking up that system of reference."

[27] Cf. the following passages from *On Certainty*: "But I could say: 'That I have two hands is an irreversible belief.' That would express the fact that I am not ready to let anything count as a disproof of this proposition." (§245) "At the foundation

if we recognize the capricious nature of *our* concept of belief.

8. Another question that poses itself besides the one about how to regard the ambiguity of the word "believe" is the question whether it is really legitimate to claim that only someone who has reached a certain level of religiosity can *understand* what it means to believe the (historical) message of Christianity. If one wishes to answer this question in Wittgenstein's sense, one will have to pay particular attention to the first words of the above-quoted passage about levels of religiosity. There Wittgenstein writes: "*In religion* it would have to be the case that corresponding to every level of religiosity ..." Presumably this means that only someone who has already stepped onto this or that rung of the ladder of religion will feel that the supposedly sense-bearing utterances belonging to levels corresponding to other rungs are senseless (or ought to count as senseless). Somebody who stands, or adopts a stance, outside religion will be able to make some sort of sense of various religious pronouncements (this is the context where those numerous books on the religions of our world become relevant). For such a person, however, the diverse rungs of the ladder of religion will have an entirely different meaning from the meaning they will have for someone who within religion stands on a particular rung of that ladder. In no way will the former person's (rough-and-ready) understanding of the ideas of one level *ipso facto* rule out the possibility of his understanding the ideas of a different level.

As we have mentioned above, Wittgenstein finds it difficult to deal with the "crown" on the head of the Bible, i.e. with the epistles of the apostles, especially with the doctrine of election by grace (predestination) as stated by Paul. According to this doctrine, God decides independently of people's conduct who will enjoy eternal salvation and who will be condemned for ever. Wittgenstein says that for him, at his level of religiosity, this principle is unintelligible and repugnant, perhaps even irreligious:

> Paul's doctrine of election by grace [...] is at my level irreligiousness, ugly nonsense. So it is not meant for me since I can only apply wrongly the picture offered me. If it is a holy and good picture, then it is so for a quite different level, where it must be applied in life quite differently than I could apply it. (C&V, 37)

It is clear that for Wittgenstein the doctrine of election by grace is not mere unintelligible "nonsense", for he calls it "ugly", and he could not characterize it thus unless he was convinced that he understood at least its rudiments.

of well-founded belief lies belief that is not founded." (§253) "If the shopkeeper wanted to investigate each of his apples without any reason, for the sake of being certain about everything, why doesn't he have to investigate the investigation? And can one talk of belief here (I mean belief as in 'religious belief', not surmise)? All psychological terms merely distract us from the thing that really matters." (§459)

Moreover, he refers to this doctrine as a "picture"; and even if he does not know how to make use of this picture, he must have seen that it really is a picture in the first place. He may try to apply the picture, but in his opinion every such application is "wrong." This, I take it, means that every application he can think of would conflict with the spirit of religion and would hence be incompatible with what (at *his* level of religion) appear to him to be essential characteristics of religion.

In spite of this difficulty Wittgenstein tries hard to render the picture of election by grace intelligible in some form. One basic problem lies in the fact that we tend to construe the notion of election in terms of reward and the notion of rejection in terms of punishment. If we do so, however, God's decision—which will be taken without taking into account the actions of elected or rejected persons—appears arbitrary and perhaps even unjust. Now, if one wishes to make the doctrine intelligible, one might try to do so by separating the two groups (of elected persons, on the one hand, and rejected persons, on the other) and regarding them independently of one another. But is it really tenable to apply the notion of punishment to the rejected people even if they could not have acted differently (in a better way)? Thus one wonders if a notion of punishment which is entirely dissociated from the circumstances of an action and the capacities of the agent can be understood. Wittgenstein claims that it would not be right to assert "God punishes, although we *cannot* act otherwise." (C&V, 87) In this case the connection between the opportunities and capacities of the agent and the verdict of his judge would be severed from the very beginning—or, rather, there would be no such connection. Therefore it is not really possible to use the word "although." What one *may* say, on the other hand, is the following:

> [H]ere there is punishment, where punishment by human beings would be impermissible. And the whole concept of 'punishment' changes here. For the old illustrations can no longer be applied, or now have to be applied quite differently. Just look at an allegory like "The Pilgrim's Progress" and see how nothing—in human terms—is right.—But isn't it right all the same? i.e. can it not be applied? Indeed, it has been applied. (C&V, 87f.)

According to this it is conceivable that a concept like that of punishment may be associated with different "illustrations" from the usual ones and lead to positive results, even though we are unable to clarify the mechanism behind this. It may well be that the notion of an "illustration" will remain opaque—but one can see why in some cases the ideas of a "different kind of life" or a "higher level of religiosity" *must* remain obscure, for otherwise one could not understand why it is possible that the expressions of a different kind of life do not "speak" to us or do not "concern" us.

In another passage Wittgenstein explores a different possibility of explaining concepts like "the punishments of hell" and "God's goodness" in a way which would not involve our human notions of punishment and goodness.

Thus one might try to tell an as it were natural-historical story about a being that divides people from the very start (that is, without taking their opportunities of action or their real deeds into account) into those who will reach a place of eternal bliss, on the one hand, and those who will be sent to a place of eternal torture, on the other. In that case neither punishment nor reward will be mentioned but only "a kind of natural law." (C&V, 92) Here everything ethical has disappeared, and "anyone to whom it is represented in such a light, could derive only despair or incredulity from it." Despair would probably be a response to thoughts about the hopelessness of one's efforts resulting from that "natural law." Incredulity would most likely be a response to the impossibility of connecting a description of this "natural-lawlikeness" in some way with ethical images (reward or punishment, leniency or strictness, etc.) or with ideas of salvation. Wittgenstein thinks that the only way of making a theory of this type intelligible to someone who has already received ethical instruction would consist in representing it "as a sort of incomprehensible mystery [*Geheimnis*]." In such a case, however, one would surely have to add what Wittgenstein writes regarding the death of the Saviour, viz. that if one wants to talk of a "mystery" here, one will first need to elucidate the grammar of the word "mystery." (DB, 220) In other words, it is *this* use of the word "mystery" that lies outside the domain of which we have a good panoramic view.

It appears that here (at *his* level of religiosity) Wittgenstein sees himself facing a riddle:

> "He has chosen them, in his goodness, and you he will punish" really makes no sense. The two halves belong to different kinds of perspective. The second half is ethical and the first not. And taken together with the first the second is absurd. (C&V, 93)

The second half is unmistakably ethical, for the notion of punishment makes no sense if one does not assume that the punished person has at least on the face of it *deserved* his fate and in some way *incurred* blame and penalty. The first half speaks of a choice which is supposed to be independent of desert and responsibility and hence unconnected with an ethical idea of reward and punishment. (The concept of goodness used in the quoted passage is hardly intelligible anyway; in no way can it be regarded as the counterpart of the severity of a person who punishes others.) Accordingly we are, as Wittgenstein suggests, dealing with a picture which—at *his* level—is unintelligible.

9. One of the important and interesting features of Wittgenstein's thought is the suggested specificity of the way religious pictures may do their job. It seems that at different levels a religious picture can fulfil completely different tasks. The fact that it has this power may be due to its intervening in different ways in the lives of the people concerned. For such a picture to be able to have this effect its parts must be in harmony and it must fit the life of a person who "passionately seizes on" it. In this respect religious pictures

are fundamentally different from pictures that are meant to decorate, to edify or to stimulate in an aesthetic way. Such roughly speaking "aesthetic" pictures need not be coherent—their parts need not be in harmony: their lack of harmony may enhance their capacity to decorate, to edify or to stimulate. Our attitude towards them is completely different from the way we approach religious pictures. Even if their "spirit" remains alien to us, we may be able to understand them in respects in which religious pictures (and perhaps ethical pictures, too) remain unintelligible to us if there is no room for them in our lives.

In a passage already quoted Wittgenstein writes that religious belief does not refer to the letter of the Gospels but to the spirit (*Geist*) of their message. A few lines further down the page he notes that it is the spirit which "puts what is essential, essential for your life, into these words." (C&V, 37) What does the word "spirit" mean here? Of course, the sense in which Wittgenstein uses the word is not particularly precise—but that is something one should not expect from a word whose ordinary-language uses form such a medley. Wittgenstein's own use of the word has certain aspects that it would be better not to overlook. Here I shall try to give a brief account of these aspects. It must be remembered that Wittgenstein does not always use the word in the same—variegated—sense. One fairly typical use of the word "spirit" can be found in the following passage:

> In the metropolitan civilization the spirit can only huddle in some corner. And yet it is not for instance atavistic and superfluous but hovers above the ashes of culture as an (eternal) witness—as if an avenger of the deity. | As if it were awaiting a new incarnation (in a new culture). (DB, 46)[28]

Here as in similar passages "spirit" stands for the spiritual element of culture as opposed to the anti-spiritual aspects of civilization. Thus the word does not refer to the spirit, or vital character, of a specific culture but to spirit in general which finds different forms of expression in different cultures. Thus understood, civilization does not have a different spirit from culture but no spirit at all.[29]

Another and equally characteristic use of the word "*Geist*" (not "spirit" but "mind") is that in which the mind of a human being is contrasted with his soul. In this sense Wittgenstein writes at one point that Frank Ramsey "had an ugly *mind*. But not an ugly soul." (DB, 8) This kind of ugliness was

[28] Cf. C&V, 5: "I once said, and perhaps rightly: The earlier culture will become a heap of rubble and finally a heap of ashes; but spirits will hover over the ashes."

[29] In this and various other passages one cannot fail to hear echoes of Spengler ("civilization" vs. "culture", etc.). As regards the relation between Wittgenstein and Spengler in general, see Joachim Schulte, *Chor und Gesetz* (Frankfurt am Main: Suhrkamp, 1990), pp. 33–35; Rudolf Haller, "Was Wittgenstein influenced by Spengler?", in Haller, *Questions on Wittgenstein* (London: Routledge, 1988), pp. 74–89.

connected with his being a "bourgeois thinker": he refused to question things in a radical way and merely wished to give shape to the existing structure of thought (C&V, 24). The kind of *Geist* intended here is something like the basic character of a man in so far as it finds expression in his attitude to spiritual (or intellectual) matters.

These two uses are to be distinguished from a third use of the word "*Geist*", which also occurs in certain passages where Wittgenstein describes his attitude towards civilization. A particularly characteristic passage is the following quotation from a draft of a preface written in 1930:

> This book is written for those who are in sympathy with the spirit in which it is written. This spirit is, I believe, different from that of the prevailing European and American civilization. The spirit of this civilization the expression of which is the industry, architecture, music, of present day fascism and socialism, is a spirit that is alien and uncongenial to the author. This is not a value judgement. (C&V, 8)

This sense is different from the first and second ones, and in this third sense our civilization too has a spirit, a quite specific spirit, which can be compared with the spirits of different civilizations and cultures. This type of comparison is possible because such a spirit has a kind of face, a certain physiognomy. This spirit finds expression, it assumes concrete form in the artefacts and practices of a given era. In saying that his response "alien and uncongenial" involves *no* value judgement Wittgenstein responds to this civilization in a way similar to a certain type of reaction to a human face. After all, it is *possible* to find a face alien and uncongenial without thereby saying anything about the person whose face it is, in particular, without claiming that he is an alien and uncongenial sort of person.

We respond to the spirit—of a culture, a style, an artist or a religion—but this kind of spirit is not immediately given to us. What we respond to is an *expression* of that spirit—a physiognomy, certain works and practices in which it manifests itself and assumes concrete shape. Our way of responding to an expression is different from responses to forms that are not perceived as expressions. Here it is important not to lose sight of the fact that our concept of expression has two sides: On the one hand, I can learn from the facial expression of another person that he is cheerful, sad or euphoric (I *simply see* that he is cheerful, etc.). On the other hand, his facial expression is one of cheerfulness, sadness or euphoria, and *this* kind of expression I can discover and recognize in the faces of other people, too, or when contemplating a work of art. Never is expression mere external form. It is possible to judge that a mountain lake or a tree is graceful, sublime or sinister; and in doing so one has at the same time attributed a certain expression to this mountain lake or that tree. This need not involve any sort of personification. But one can see in which way personification can get going from here.[30]

The spirit of an era or a culture finds expression in its artefacts and practices. One method of rendering an era or a culture intelligible is by putting together artefacts and practices in such a way that they form a physiognomy which is characteristic of that spirit. This spirit may be strange or unsympathetic. But if this is your reaction, you cannot really claim that you do not understand it.[31] A face may appear reserved; but if I articulate this impression, I do not say that it expresses *nothing*. On the contrary, by articulating my impression this way I ascribe a certain expression to that face—this is how it has struck me.

The spirit of religious doctrines or ceremonies may be characterized by emphasizing certain physiognomic traits. Here the physiognomy is an expression of their spirit. Once you have identified the outlines of the physiognomy of this spirit, you will develop a feeling that tells you which expression will fit this spirit. In such a case one may speak of gestures that correspond or fail to correspond to the spirit of a certain religion. For the spirit itself is as it were a gesture; it finds expression in a gesture.[32]

10. These and similar considerations mould many reflections of Wittgenstein's, among them his observations on religious questions and concepts. A particularly relevant set of remarks concerns our understanding of divine miracles. In a diary entry of 1931 he writes:

> If one wants to understand as Dostoyevsky[33] did the miracles [*Wunder*] of Christ such as the miracle at the wedding of Cana, one must consider them symbols. The transformation of water into wine is astounding at best and we would gaze in amazement at the one who could do it, but no more. It therefore cannot be what is magnificent.—[...] It must be the marvelous [*das Wunderbare*] that gives this action content and meaning. And by that I don't mean the extraordinary or the unprecedented but the spirit in which it is done and for which the transformation of water into wine is only a symbol (as it were) a gesture. A gesture which (of course) can only be made by the one

---

30 See Fr, 237: "The idea that one can summon an inanimate object to oneself as one can summon a person. Here the principle is that of personification." Cf. Fr, 239: "Personification will, of course, play a large role in these simple pictures [...]."

31 Here the word "understand" is used in a very basic sense. In this sense I may understand what there is to be seen in Titian's painting "Venus with Organ Player" without understanding why a naked woman and a fully dressed organ player plus organ have come together in a delightful garden.

32 See DB, 84f.: "When I read in a fairy tale that the witch transforms a human being into a wild animal, it is also the spirit of this action, after all, that makes an impression upon me."

33 *Brothers Karamazov*, part 3, book 7, ch. 4 (editor's note re DB, 82).

> who can do this extraordinary thing. The miracle must be understood as a gesture, as expression if it is to speak to us. I could also say: It is a miracle only when *he* does it who does it in a marvelous spirit. Without this spirit it is only an extraordinarily strange fact. I must, as it were, know the person already before I can say that it is a miracle. I must read the whole of it already in the right spirit in order to sense the miracle in it. (DB, 82–84)

To understand a miracle rightly, i.e. symbolically, presupposes seeing it as a gesture. In this way one can understand the spirit in the context of which an extraordinary act can appear as a miracle in the first place. As Wittgenstein emphasizes, primitive religions and magical rituals[34] too involve symbolic achievements: "And magic is always based on the idea of symbolism and language." (Fr, 237)[35] In many cases those practices can be interpreted as gestures: "In the ancient rites we have the use of an extremely developed gesture-language." (Fr, 242)

One can understand such gestures without being touched by them: they need not make a great impression. A man who stands at a different religious level from Aliosha in Dostoyevsky's *Brothers Karamazov* may well be able to imagine what it would be like to conceive an extraordinary occurrence as a symbolic move, as a divine gesture, and hence as a miracle. But if I am to be able to imagine that, I myself need not regard that occurrence as a miracle. It is this relation between my understanding a miracle as a gesture and my simultaneous incapacity to grasp the miraculousness of a miraculous event which Wittgenstein characterizes in the following passage:

> A miracle is, as it were, a *gesture* which God makes. As a man sits quietly and then makes an impressive gesture, God lets the world run on smoothly and then accompanies the words of a saint by a symbolic occurrence, a gesture of nature. It would be an instance if, when a saint has spoken, the trees around him bowed, as if in reverence.—Now, do I believe that this happens? I don't.
>
> The only way for me to believe in a miracle in this sense would be to be *impressed* by an occurrence in this particular way. So that I should

[34] In his remarks on Frazer's *Golden Bough* Wittgenstein makes no general distinction between religious and magical practices. Perhaps this is in deliberate opposition to Frazer.

[35] Peter Hacker regards the quoted sentence as a dogmatic statement and says that it "is either a stipulative definition or an overhasty generalization" ("Developmental Hypotheses and Perspicuous Representations: Wittgenstein on Frazer's *Golden Bough*", in P. M. S. Hacker, *Wittgenstein: Connections and Controversies* (Oxford: Clarendon Press, 2001), pp. 74–97, quotation p. 80). In my view that sentence claims much less. I think it should be read in the following sense: In studying magical actions we again and again encounter elements that can be interpreted symbolically, i.e. in terms of linguistic representation ("standing for") and acts performed by linguistic means.

> say e.g.: "It was *impossible* to see these trees and not to feel that they were responding to the words." Just as I might say "It is impossible to see the face of this dog and not to see that he is alert and full of attention to what his master is doing." And I can imagine that the mere report of the *words* and life of a saint can make someone believe the reports that the trees bowed. But I am not so impressed.[36]

Someone who sees the way the spirit of a religious message or a ceremony finds expression and discovers how it can be interpreted as a gesture or what kind of gesture would fit it could be said to understand the sense of that message or ceremony. Whether this sense concerns him—whether he can make use of it—is a different question. In this context Wittgenstein plays with the idea that what religion says by means of words might entirely be reduced to gestures:

> One could conceive a world where the religious people are distinguished from the irreligious ones only in that the former were walking with their gaze turned upwards while the others looked straight ahead. And here the upward gaze is really related to one of our religious gestures, but that is not essential and it could be the other way round with the religious people looking straight ahead etc. What I mean is that in this case religiosity would not seem to be expressed in words at all and these gestures would still say as much and as little as the words of our religious writings.[37]

This imaginary case serves to make particularly clear the extent to which the religiosity of gestures (or words) depends on the role these gestures (or words) play in the lives of the people in question. Often a person who wishes to get across a religious message cannot do much except for describing certain events,[38] tell a story, make a gesture—"whatever gloss someone may want to put on it!" (C&V, 32) If a person can make something of a certain gesture,

[36] C&V, 51f., cf. C&V, 57: "The purely corporeal can be uncanny. Compare the way angels and devils are portrayed. What one calls 'miracles' must be connected with this. It must be as it were a *sacred gesture*." For the notion of a miracle, cf. LoE, 43f.

[37] DB, 61. Cf. DB, 207f.: "Christianity says: Here (in this world)—so to speak—you should not be *sitting* but *going*. You must away from here, and should not suddenly be torn away, but be dead when your body dies. | The question is: How do you go through this life?—(Or: Let this *be* your question!)—Since my work, for example, is only a sitting in the world, after all. But I am supposed to go and not just sit."

[38] Cf. Fr, 236: "Here [regarding the life of the King of the Wood of Nemi] one can only *describe* and say: this is what human life is like." In a particularly agitated mood Wittgenstein notes in his diary (DB, 183f.): "Call it a sickness! What have you said by that? *Nothing.* | Don't *explain*!—DESCRIBE! Submit your heart and don't be *mad* [*bös*] that you must suffer so! This is the advice I should be giving myself. When you are sick, accommodate yourself to the sickness; don't be mad that you are sick."

he may be able to make it part of his repertoire of religious ideas. If he wants to put a gloss on it which expresses these ideas, then he will have understood it as a religious gesture.[39]

With respect to Frazer's account of the life of the King of the Wood of Nemi Wittgenstein writes: "One would like to say: This and that incident have taken place; laugh, if you can." (Fr, 236) Of course, this invitation itself—"Laugh, if you can!" (cf. "whatever gloss someone may want to put on it!")—is a gesture.[40] If here someone feels like laughing, he may still understand that what he has before him is a representation of "the majesty of death", as Wittgenstein points out, but this representation fails to touch him—the sinister and frightening aspects of this thought do not "mean" anything to him. In this vein Wittgenstein continues:

> The religious actions, or the religious life, of the priest-king are no different in kind from any genuinely religious action of today, for example, a confession of sins. This, too, admits of being '*explained*' and not explained. (Fr, 236)

Most people will be able to recount the story, to make corresponding gestures, to describe the relevant context and in this sense to "explain" what is going on in the cases of the priest-king or of confessing sins. But what it means to conceive these actions as religious ones is something which no one who is blind for their religious aspects will be able to explain.[41]

11. In saying that the religion Jukundus expresses by his words is really the same as the one expressed by Job through exclaiming "The Lord hath given, the Lord hath taken away; blessed be the name of the Lord!" Wittgenstein means—and this is my thesis—that the spirit of these expressions is the same. Both expressions are gestures of the same kind. In other words: as regards *both* expressions, it is possible to demonstrate by means of one and the same gesture that one has understood them.

Moreover, the spirit shared by Jukundus's and Job's religions is a spirit Wittgenstein is in sympathy with. It is the spirit of acceptance and calm resignation.[42] This spirit corresponds to Wittgenstein's level of religiosity. In March 1937 he observes in his diary: "If you want to quarrel with God, that means that you have a false concept of God. You are superstitious. You have

39 Cf. the passage quoted above in §5 on religious images (DB, 173).

40 Cf. footnote 19 above on requesting someone to take a stance.

41 See Hacker, "Developmental Hypotheses", 76f., 91; Frank Cioffi, "Wittgenstein and Obscurantism", in Cioffi, *Wittgenstein on Freud and Frazer* (Cambridge: Cambridge University Press, 1998), p. 183, especially pp. 186–205. See below, Appendix, note 2.

42 "Thoughts at peace. That is the goal someone who philosophizes longs for." (C&V, 50) Cf. the second quotation in footnote 38 and, of course, PI, §133. See my "Wittgenstein's Quietism", in *Metaphysics in the Postmetaphysical Age*, ed. Uwe Meixner (Wien: Hölder-Pichler-Tempsky, 2001), pp. 37–50, especially §7.

an incorrect concept when you get angry with fate. You should rearrange your concepts. Contentment with your fate ought to be the first command of wisdom."[43] But the resigned gesture of both Jukundus and Job agrees not only with a religious strand of Wittgenstein's thought but also with an essential trait of his philosophy. This gesture is in accord with the spirit which finds expression in remarks such as PI, §654[44] and in Wittgenstein's articulation of the yearning for philosophical peace (see e.g. PI, §133).

The gestures Wittgenstein likes are simple gestures. In opposition to such gestures stand the ostentatious, perhaps even pretentious gestures of the epistles of Paul—the "crown" on the head of the Bible:

> In the Gospels—as it seems to me—everything is *less pretentious*, humbler, simpler. There you find huts;—with Paul a church. There all human beings are equal and God himself is a human being; with Paul there is already something like a hierarchy; honours, and official positions.—That is, as it were, what my NOSE tells me. (C&V, 35)

What Wittgenstein finds especially problematic is Paul's doctrine of election by grace. If one follows the complicated considerations by means of which Wittgenstein attempts to make sense of this doctrine, one may arrive at the same conclusion reached by him when he says that this is a brandishing of words; here we find no simple gesture; I am not shown why I should not remain seated but go.

Appendix

Note 1: In his remarks on fire rituals, and in particular on the Beltane Festival (Fr, 246ff.), Wittgenstein again and again speaks of the "depth" of these practices and their contemplation, and in this connection he often uses the word "sinister." In this context Wittgenstein makes a claim which is not easy to interpret. A correct reading of this passage, however, is important for it is of fundamental significance. Wittgenstein writes:

> Indeed, how is it that in general human sacrifice is so deep and sinister? For is it only the suffering of the victim that makes this impression on us? There are illnesses of all kinds which are connected with just as much suffering, *nevertheless* they do not call forth this impression. No, the deep and the sinister do not become apparent merely by our coming to know the history of the external action, rather it is *we* who ascribe them [*tragen es wieder hinein*] from an inner experience. (Fr, 249)

[43] DB, 217f. It is clear that at this point no opposition between the passion of religiosity and the "coldness" of wisdom is intended.

[44] "Our mistake is to look for an explanation where we ought to see the facts as 'proto-phenomena'. That is, where we ought to say: *this language-game is played.*"

If this passage is to be connected in a coherent fashion with Wittgenstein's other thoughts, it must not be construed in a way that makes it appear as if what is deep and sinister is a more or less typical feeling which is triggered in us by such practices (or reports of them) and is then projected by us onto those very practices.

I suspect that the "inner experience" [*Erfahrung in unserm Innern*] mentioned by Wittgenstein in this passage is the extremely complicated and indirect insight that certain ways of action correspond to certain patterns which force themselves on us in an almost irresistible manner. Someone who contemplates the structure of what seems at first glance an innocent practice and suddenly notices how naturally it comes to him to think "At this point a human sacrifice is required" (or "Without a human sacrifice this entire procedure is senseless" etc.) may be dismayed by this thought. At this moment he becomes aware of what is sinister and deep about it—his dismay brings forth what is sinister, and the inexorability of the image forcing itself upon him suggests depth.

Such a pattern is something that resides in ourselves; it is something natural, it conforms to our nature; it is *deeply* rooted within us. This state of things lies in darkness (Wittgenstein's German word "*finster*" means both "dark" and "sinister"). The fact noted by Wittgenstein that we tend to "carry back" sinisterness and depth into a described or observed practice does not mean that we tend to project a certain feeling onto it and thus confer sinister and deep aspects on it. What it means is that we (can learn to) see it as something sinister and deep, just as we (can learn to) hear a certain sequence of notes as a melody, as the inversion of a given theme, etc. If a practice is seen according to a pattern which in its turn corresponds to a pattern of our own nature, it assumes a certain expression—surely a deep and perhaps a sinister one.

When confronted with certain rituals we often cannot help exclaiming words to the effect that "This practice is *obviously* ancient." (Fr, 248) This sort of response does not originate in an historical hypothesis; it is a matter of instinct, an expression of our spontaneous recognition of a certain pattern which belongs to our nature. This too is "a document of a tendency in the human mind." (LoE, 44) It is an expression of what Peter Hacker aptly calls "the common *wonder* of mankind"[45]: we marvel at the patterns of human life; and the fact that people marvel at certain kinds of patterns is a fact whose recognition helps human beings to recognize themselves and each other as beings of a certain kind. It is a reasonable move on Hacker's part to associate this common wonder with *Philosophical Investigations*, §206, a passage of central importance in Wittgenstein's work: "The common behaviour of mankind is the system of reference by means of which we interpret an unknown lan-

[45] As regards this common kind of wonder Hacker writes: "[T]he forms of impulse, symbolic and expressive, to which it [the common wonder of mankind] gives rise in us is the point of reference for rendering intelligible the meaning of magical rites of primitive societies."

guage." If one reads this passage in the light of the above considerations one can hardly help supposing that what Wittgenstein is talking about is a common human nature.[46]

Note 2: I read the quoted passage "This, too, admits of being 'explained' and not explained" (Fr, 236) in a way which is different from the interpretation given by Cioffi and Hacker. These authors read the first occurrence of "explained" (italicised[47] and in quotation marks) in such a way that the word receives a completely different sense from that given to the second occurrence of "explained" (without quotation marks). In its first sense "explain" is said to mean "elucidate", "clarify", etc. in the sense of a hermeneutic kind of understanding. In its second sense it is said to mean scientific (causal, hypothetico-deductive, strictly historical) explanation.

Of course, this is a possible (admissible) interpretation of the passage quoted, but I feel that it is not consonant with Wittgenstein's intentions. I think that both times the word "*erklären*" is used in the same sense. The point of the passage is that in this context explanation does not result in what it is expected to result in, viz. satisfaction, "understanding." Take the explanation of a given passage from a piece of music. I may inform another person about harmonic progressions, rhythmic effects, parallels in other works, and so on and so forth. By telling him about these things I have certainly conveyed some information; and in a sense this information may even be "exhaustive." But if the other person—in accordance with the silent premise of the quoted passage from Wittgenstein—has no or too little musical knowledge, my information will not speak to him, nor will it tell him anything.

In the same way one may tell a person willing to learn about these matters many things about religious practices—but if he has no religious bent, i.e. if those gestures do not speak to him, that sort of explanation will achieve next to nothing. Such explanations will remain idle. The learner may understand the letter, but he will not grasp the spirit. The letter may be explained, the spirit does not admit of explanation. Thus the quoted passage from Wittgenstein does not involve an ambiguity afflicting the word "explain." The point is that we are dealing with two different *explananda*—religious acts as historical occurrences [LETTER] vs. religious acts as part of a practice into which one may be initiated and which needs to be lived to be understood [SPIRIT]. It may be quite possible to explain a religious act, but explaining it will not achieve much if the practice concerned (e.g. confession of sins) does not speak to me. That does not mean that this gesture will have to remain alien to me forever. There are all kinds of means (practicing, training, exercises)

---

46 Eike von Savigny argues against this view, see his "Viele gemeinsame menschliche Handlungsweisen", in v. Savigny, *Der Mensch als Mitmensch* (München: dtv, 1996).

47 In Wittgenstein's manuscript the first occurrence of "explained" ("*erklären*") is between quotation marks, but the word is not underlined. Only in the later typescript version is the word between quotation marks also spaced out.

which may help me to reach understanding. And once I have been trained that way, explanations too may be helpful. Until this point has been reached explanations will leave me cold; they do not concern me, they do not ring a bell. In such a case I shall for instance remain incapable of hitting unaided on illuminating parallels that would mean something to people who are sensitive to religious issues.

To be sure, the ethnological approach favoured by Wittgenstein suggests that we look at matters from a detached (objective) point of view, but at the same time understanding would be impossible unless we are involved to such an extent that the things observed (practices, ceremonies, rituals, etc.) mean something to us. That is, we must be able to find a continuation without outside assistance. To revert to our musical parallel: We must be able to play (sing, whistle) a passage the way it is or may be meant.[48]

[48] This paper was written a number of years ago. If I wanted to write a new version of it, I would certainly wish to take into account several articles and two or three books that have been published in the meantime. Above all, I would signal more recent editions of some of my sources. Here, I just want to mention the most important ones. Alfred Nordmann's translation of *Denkbewegungen* (mentioned in note 3, above) has appeared in Ludwig Wittgenstein: *Public and Private Occasions*, edited by James C. Klagge and Alfred Normann, Rowman & Littlefield, 2003. A new edition, revised by Ilse Somavilla and Brian McGuinness, of Paul Engelmann's memoir of and correspondence with Wittgenstein (note 7) has come out in 2006: *Wittgenstein - Engelmann: Briefe, Begegnungen, Erinnerungen*, Innsbruck: Haymon. A completely revised and much enlarged edition of Wittgenstein's correspondence with his Cambridge friends (note 5) has been brought out by Brian McGuinness: *Wittgenstein in Cambridge* (Oxford: Blackwell, 2008).

FSC
www.fsc.org
MIX
Papier | Fördert gute Waldnutzung
FSC® C083411

Kolibri 360 GmbH
Ferdinand-Jühlke-Straße 7
99095 Erfurt, Deutschland
produktsicherheit@kolibri360.de